T0321001

Communications, Navigation, Sensing and Services

RIVER PUBLISHERS SERIES IN COMMUNICATIONS

Consulting Series Editors

MARINA RUGGIERI
University of Roma "Tor Vergata"
Italy

HOMAYOUN NIKOOKAR
Delft University of Technology
The Netherlands

This series focuses on communications science and technology. This includes the theory and use of systems involving all terminals, computers, and information processors; wired and wireless networks; and network layouts, procontentsols, architectures, and implementations.

Furthermore, developments toward new market demands in systems, products, and technologies such as personal communications services, multimedia systems, enterprise networks, and optical communications systems.

- Wireless Communications
- Networks
- Security
- Antennas & Propagation
- Microwaves
- Software Defined Radio

For a list of other books in this series, please visit www.riverpublishers.com

Communications, Navigation, Sensing and Services

Editors

Em. Prof. dr. ir. L. P. Ligthart

Delft University of Technology, NL

Prof. dr. R. Prasad

CTIF, Aalborg University, Denmark

River Publishers

Aalborg

Published, sold and distributed by:
River Publishers
PO box 1657
Algade 42
9000 Aalborg
Denmark
Tel.: +4536953197

www.riverpublishers.com

ISBN: 978-87-92982-39-1
© 2013 River Publishers

Preface

Five years ago I discussed with Prof Ramjee Prasad our more than 40 years exciting experiences in mobile and wireless communications and the overwhelming and sometimes unforeseeable growth in this field. In our discussion we faced back to the introduction of GSM and the impact it had for "young" and "old" in Society. Nowadays we have 3G and 4G mobile communications networks supporting an incredible number of "aps". We do not feel comfortable when we are not connected to the "world" via our "personal" devices. Prime question remained in our meeting: "Is it possible that future radio has a similar impact in Society?"

We wondered if such big changes as we have seen in existing generations of mobile and wireless communication are feasible nowadays. Present mass production and mass utilization of advanced technologies, systems, networks and applications have demonstrated an impressive record in world economy with an important series of commodity products. Trying to answer the just raised question necessitates a plan for needed, self-learning and most intelligent applications that has a potential big impact on Society and which requires the development of a series of new mass products. In order to receive support from societal organizations the new products should come into use within 10 to 20 years from now. We have the vision that in the 2020tees self-learning systems which fully integrate the fundaments of communication, navigation, sensing and services will play a role in daily life in society.

Three years ago we decided to give high priority to our vision by establishing a non-profit foundation based on our visionary plans. We informed many colleagues and received highly positive responses to continue our ideas for establishing such a new foundation. This resulted in the establishment of the Conasense foundation at the notary office in Delft, The Netherlands on

November 8, 2012. We were happy to have Prof. M. Ruggieri from University of Rome, Tor Vergata and dr. H. Nikookar from Delft University of Technology in the board of the new foundation.

In February 2012 the first annual Conasense workshop was organized in Amsterdam. Contributors were: Prof. Leo Ligthart, Dr. Homayoun Nikookar, Prof. Ramjee Prasad, Em Prof. Durk van Willigen, Dr. Tomasso Rossi, Prof. Ole Lauridsen, Dr. Silvester Heijnen, Prof. Kwan-Cheng Chen, Prof. Mehmet Safak, Prof. E. Fledderus, Prof. Silvano Populin, Prof. Enrico Del Re, Prof. Vladimir Poulkov, Rajeev Prasad.

Without exaggeration I state that the workshop was extremely successful. There were 13 contributions from research organisations all over the world. Nine contributions laid the fundament of the Chapters of the first Conasense book.

On the one side, I thank the authors and participants of the first annual Conasense workshop for their contributions and for laying the fundaments of a big change.

On the other side, I hope that readers of the book are inspired by the topics that need future research and developments and that they become motivated to work on those Conasense topics and finally that they are eager to join Conasense community.

Leo Ligthart
Chairman Conasense
www.conasense.org

Contents

1

CONASENSE: A New Initiative on COmmunication, NAvigation, SENsing and SErvices

Leo P. Ligthart

Chairman Conasense

1.1 Introduction

Motivation of this initiative is the key issue of this first chapter. In this introductory section some backgrounds are given which have led to this initiative, followed by the contents of this chapter.

In the last decade wireless technology and systems created an explosive growth in personal and group applications which offer a wide range of data services. Services started with voice, video and image transmissions. Bottleneck in the beginning was often the limited data transmission capacity. However, breakthroughs in throughput realized in the recent past or expected in near future allow for much higher data rates over the same wireless channel and thus for support of many more demanding services.

Nowadays (nearly) all devices can be connected into a wireless network and execute ad-hoc and/or specific actions. This leads to new market initiatives in which the number of services becomes manifold and more diverse.

COmmunications- NAvigation-SENsing-SErvices (CONASENSE), 1–17.

Present mobile systems have ample opportunities to perform optimally in 'real-time', in an energy-efficient way by taking their operational environment into account.

Some challenging multi-disciplinary research programs, in which wireless networking play a large role, have shown that those programs have communications, navigation, sensing and services in common.

In 2009 the author launched the initiative for a Scientific Society on Communication, Navigation, Sensing and Services (CNSS). Prime objective is to come with a long-term vision on CNSS, 20 to 50 years from now, and supplemented with action plans needed for reaching this objective. The initiative has been strongly supported by Prof. Dr. Ramjee Prasad from Aalborg University and Prof. Dr Marina Rugieri (MR) from Tore V. University Roma, who was IEEE President AES at that time. The proposed and approved name of the new Society became CONASENSE in 2011. Prof. Ligthart has accepted the Chairmanship position and Prof's Prasad and Rugieri have become Board members of CONASENSE. Dr. Homayoun Nikookar from Delft University is the first Secretary. More board members will be appointed in 2013.

In this chapter the initiative is worked out. After this introduction (Section 1) the chapter has two more sections. In Section 2 some examples of past programs and projects executed at my former university are described to illustrate the importance of CONASENSE. Its characteristics are explained in Section 3 and include its vision and important aspects in the start-up phase as discussed during the CONASENSE kick-off meeting in February 2012 with invited representatives from academia, research institutes and industry. Section 3 ends with the organizational structure for CONASENSE workshops. The end of this chapter gives major conclusions.

1.2 Examples Illustrating CONASENSE Importance

As said in the introduction many multi-disciplinary research programs and projects started world-wide in the last decade. Most research topics concern application areas which receive high priority in Society such as security, health and environment. Mission requirements for these applications necessitate breakthrough research in which satisfying of accuracy and reliability aspects as well as real-time operation of multiple CONASENSE systems in a network is crucial. It was and is the conviction of the author that the outcome of this research has a huge impact on the social well-being of people and finally on the overall functioning of local, national, international and global organizations and thus of Society. For this reason first ideas on CONASENSE existed already in 2005. During that time several projects with definitive CONASENSE features were in progress under my supervision as director of the International Research Center for Telecommunications and Radar (IRCTR) of Delft University of Technology, The Netherlands.

Below two of my projects are summarized as examples illustrating different application areas for niche markets in which CONASENSE aims and objectives can be explained.

1.2.1 Broadband Services with Localization Functionalities in not-yet Existing Maritime Applications [1]

Based on 3G and 4G wireless communications prospects, a new maritime CONASENSE system was proposed offering a wide variety of accurate and reliable services based on high data rate transmissions over distances up to 20 kilometers. Special attention is paid to an operational availability of more than 99 percent, independent on actual system conditions.

With this new maritime system the skippers can make use of broadband and/or shared services important for doing their business or having entertainment on ships. The system not only provides faster web surfing and quicker file downloads on board, but also enables several multimedia applications, such as real-time audio and video streaming, multimedia conferencing, high-definition TV (HDTV), video-on-demand (VoD) and interactive gaming, as shown in Figure 1.1.

On the one hand, a guaranteed quality of broadband services such as voice-over-IP (VOIP), video images, Internet access and IPTV attracts new users. On the other hand, a large coverage area up to 20 km means that the new maritime system needs fewer infrastructures than terrestrial systems and consequently has fewer costs. On top of this, the facility to get high resolution localization information of various users offers new services as emergency services after a fire onboard of a vessel as visualized in Figure 1.2.

Any ship that encounters a disaster is required to send its SOS signal, live video and all sensors data to the base station BS. The high data rate ensures that streaming video and large amount of data from sensors arrive at the BS

Broadband Services   Localization

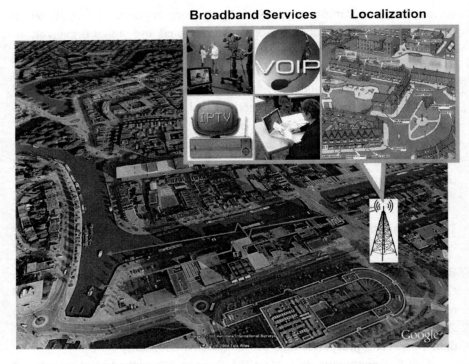

Fig. 1.1 CONASENSE example 1: Near to medium range services for maritime users.

in time, which is essential for providing optimal emergency services after maritime accidents. On shore, experts in the control room are able to analyze the cause of the tragedy, decide the best rescuing operation, and monitor the accident scene. At last, several wrecking ships are assigned to the live spot with the determined rescuing operation to help the ship in trouble.

A major system feature concerns the use of green radio techniques that enhances the quality of service with less energy consuming and less hardware on board of vessels compared to existing systems. Users of the maritime system will experience this feature as positive. It is also appreciated that the system offers versatile broadband wireless services and may significantly increase the safety of water transportation.

A major technological challenge concerns the use of a fast broadband access technology that enables flexible bandwidth, fast link adaptation and positioning capability. By employing smart radio concepts the actual adaptation to optimize operational system conditions in combination with intelligent

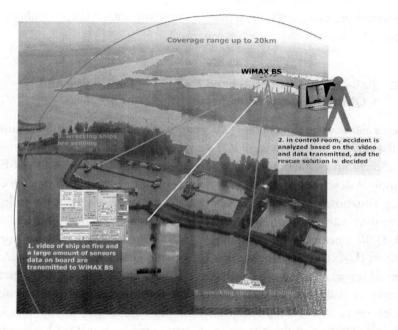

Fig. 1.2 CONASENSE example 1: Near to medium range emergency services provided by the new maritime system.

radio resources allocation can be realized. Adaptive modulation and coding schemes allow for assigning the best available radio spectrum and required transmit power based on the understanding of the actual wireless channel. Minimization of the transmit power and high resolution positioning can be realized by an advanced antenna system of the base station with adaptive beam forming capabilities. Multiple users build up a cooperative network allowing for distributed beam forming in order to guarantee reliable transmissions to most far users at the edge of the coverage area.

Adaptive modulation is meant to combat the time-selective and non-deterministic fading of wireless channels and to optimize the power and throughput conditions. Various modulation parameters, i.e., modulation levels, coding rate, multiple access and multiplexing schemes, can be changed in accordance with the instantaneous channel situation to get a higher throughput and/or better transmission quality. When the actual channel leads to a high signal to noise ratio at receive higher modulation parameters are assigned to get a higher throughput; when the channel quality is poor, lower modulation parameters are assigned to guarantee a minimum transmission quality of the system [2].

1.2.2   Emergency Ultra-wideband RadiO for Positioning and COMmunications (Europcom) [3]

When a catastrophe (for example a big explosion, an earthquake, a tsunami or a big fire) takes place permanent ICT services needed by rescue teams are often insufficient and sometimes major services can be destroyed. CONASENSE example 2 describes a project how services can be established in such emergency situations. In the 6th framework EU-project Europcom the following consortium partners participate: Thales Research & Technology (UK) Limited, Delft University of Technology, IMST GmbH (FRG), Thales Security Systems (UK) Ltd. and Technische Universität Graz (AU). In this project the selected scenario is that after a catastrophe large buildings may be partially or completely collapsed. In such case on-line communications with rescue personnel can be difficult particularly within those buildings. In addition, safety and co-ordination of the operations is hampered by a lack of knowledge of the whereabouts of emergency staff. The project intention is to investigate and demonstrate the use of Ultra-Wide-Band (UWB) radio technology, to improve the reliability of communications and simultaneously allow the precise location of personnel to be displayed in a central control vehicle.

The extremely high bandwidth of UWB enables very accurate timing information to be resolved. This enables highly accurate range measurements and hence highly accurate position information, using multiple ranges from different sources. In addition, the precise timing resolution allows separation of the high levels of multipath propagation found within buildings; this permits high accuracy positioning to be maintained within buildings and also offers the potential of robust communications. UWB radar could also be used to search for survivors buried beneath rubble.

The overall objective of the project was to demonstrate the feasibility of a system for providing an emergency control centre with the positions of all emergency personnel. In addition, position information may be supplied directly to individual emergency workers and a backup robust communications system will be provided for data transfer such as sensor information to and from each emergency worker. The actual data rate to be provided

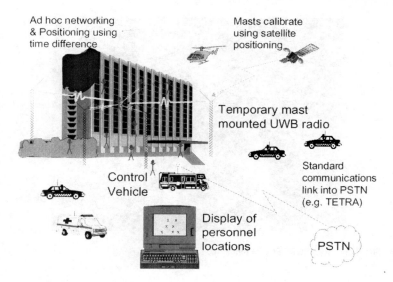

Ad hoc networking & Positioning using time difference

Masts calibrate using satellite positioning

Temporary mast mounted UWB radio

Control Vehicle

Standard communications link into PSTN (e.g. TETRA)

Display of personnel locations

PSTN

Fig. 1.3 CONASENSE example 2: UWB emergency system concept (courtesy Europcom consortium).

depends on the particular environment and number of network nodes but at least, several dozens of kbps should be available. The transferred sensor information includes the status of the particular emergency at each location (such as progress of a fire), health status, and quality of the radio link to the closest neighbor nodes. The sensors allow also for detection and localization of fire victims and generation of a 3D map of the environment. A schematic illustration of the system concept is shown in the figure.

New mobile UWB units (types: ranger, transponder and drop units for building up a dense ad-hoc network) have been developed. Ranging accuracy has been investigated in several situations: long corridors and pathways, rooms at the same floor or at different floors, stairs, cellar with thick concrete walls, etc.

In non-line-of sight cases with strong multi-path the rms deviation of the estimated range increases from less than 1cm (in free space) to 25cm or more. Transmissions through (multiple) thick walls and floors have learned that not only ranging but even network synchronization of the ranger and transponder becomes difficult.

In 2008 a demonstrator system was tested by end users (fire fighters, police organization, and other emergency services) at the premises of Fire College of London and at the Fire Fighter facilities in Rome.

Fig. 1.4   Fire fighter with EUROPCOM unit at the left shoulder.

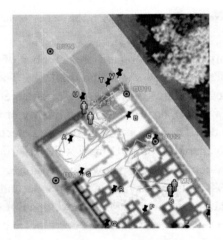

Fig. 1.5   Control centre display with trajectories of two groups of fire fighters.

A positioning performance was achieved within the target of 1m accuracy in 2D, provided that a sufficient number of drop units were deployed. In 3D, the system did provide a similar accuracy in positioning, provided that all units had also sufficient height spread. Units were carried by personnel using jackets with shoulder attachments. Movement of the personnel was tracked on the display of the Control Centre.

1.3 CONASENSE Characteristics and Structure

The new scientific platform for Communication, Navigation, Sensing and Services (acronym CONASENSE) has been announced in March, 2012 and will be established as scientific society before September, 2012. Prime aim is the formulation of the vision on how CONASENSE can solve societal problems and how Mankind can benefit most optimally from this vision. Focus is on those problems which can be solved by using new insights obtained from breakthroughs in science and technology and which integrate communications, navigation, sensing and services. Main objective is therefore to give a CONASENSE vision at time scales of 20 to 50 years from now. This vision should be based on scientific and technological challenges with substantial impact on Society. After my intensive discussion with Prof. Dr. R. Prasad we decided to put the concept of CONASENSE consisting of "The Integrated Vision" and "Benefits to Communications, Navigation and Sensing systems plus their Services" on the new CONASENSE website: www.conasense.org.

In the initial phase of CONASENSE large attention is paid to organize during 2012 and 2013 brain-storming workshops in which experts and specialists from universities, institutes and industry with knowledge in one or more areas of CONASENSE can bring in their points of view. Those experts and specialists can become full member of CONASENSE.

Workshop participants can come from all over the world. The author has the opinion that the workshops should primarily be attended by top-specialists and top-PhD's and preferable with a recommendation from one or more full members. The workshops are important for developed and developing countries, because most developing countries now should be developed countries in a time frame of 20 till 50 years from now. A first meeting, which was also the kick-off meeting of CONASENSE, was held Amsterdam at February, 2012.

Support from the International Gauss Research Foundation (GRF), with Chairman Prof. Dr. D. van Willigen and with its address in The Netherlands, is highly acknowledged. A selected group was invited for this meeting. The participants are listed in Appendix I.

The kick-off meeting has been planned in order to discuss topics which should have an output important for Science, Politics (based on the idea that decision makers require CONASENSE-related multi-disciplinary advices), Scale of Economics and Impact on Social Life (based on the idea that

Society becomes more and more dependent on CONASENSE facilities). Future meetings can be organized in 3 parts: part 1 as brainstorm workshop, part 2 as symposium in which decision makers attend also and part 3 as Global CONASENSE Forum resulting into major future policies and strategies between decision makers and CONASENSE.

A major challenge for those future meetings will become the motivation and inherently the set up of a roadmap for new CONASENSE business opportunities in future (5, 20, 50 years) scenarios illustrated with talked-about examples and ways to apply new capabilities. The examples should at least illustrate some future 'daily-life' situations in which CONASENSE may play an obvious role. The examples can be best presented in interactive "video-type" computer simulations.

CONASENSE has been established as a Foundation in The Netherlands. An important action for the Foundation is to make agreements with various universities, institutes, organizations and companies, active in the field.

In Section 1.3.1 the discussions during the kick-off meeting are summarized, illustrating how the CONASENSE characteristics are judged by the participants.

1.3.1 Summary of the Kick-off Meeting

Before the kick-off event was held the invitees were requested to bring in topics and visionary presentations during the meeting. From the responses of all invitees it became obvious that a very interesting program could be scheduled and that future use and substantial impact of various modalities of CONASENSE systems active from ground, sea, air and space are foreseen. The high-quality material used in the kick-off meeting has resulted into 2 prime actions.

(a) The Chairman will write a short document about CONASENSE essentials to be sent to various universities, institutes, organizations and companies, active in the field. The document is important for getting enthusiasm among potential partners.

(b) All presentations are worked out as Chapters 3 till 12 of the first CONSENSE book. Chapter 1 (this chapter) describes the basic characteristics. In Chapter 2 Prof. Dr. R. Prasad gives major expectations and objectives for CONASENSE. All other chapters have been written based on the individual presentations.

In the meeting 3 important themes have received attention. They are worked out in the next sections.

1.3.1.1 Most promising CONASENSE services to be available in 5–20 years

(a) It was noted that the trend is more toward an information society in which applications and services become equally important. Examples were given on more diverse and/or new services integrated in applications for air-, road- and vessel traffic control and management.

(b) Computing and Communications should be integrated. Computing can sometimes be less precise, which means that in many cases energy can be saved. An example can be robot-like operations which should not always mean precise computing for positioning. In this connection we can think about lowering energy consumption in case the swarm problem in Machine to Machine applications and services is investigated. We see a new trend from mobile-computing to distributed-computing channels.

(c) Software defined radio combined with cognitive radio concepts become increasingly important for new developments. After problems in processing for optimizing cognitive radio transmission have been solved more advanced and demanding new services can be realized.

(d) CONASENSE brainstorming may lead to futuristic challenges, developments, applications and services, assuming that technological problems of today and to-morrow can be solved. However, remaining technological constraints are still important such as integration of intelligent systems and their interferences.

(e) CONASENSE should have many activities as lowering energy consumption in devices, increasing functionalities of the devices, systems which operate at various frequencies and/or using frequency hopping etc.

Focus in CONASENSE is therefore essential

1.3.1.2   CONASENSE focus

Basic assumption is that CONASENSE executes multi-disciplinary activities complementary to ongoing R&D elsewhere. Focus should not be on the 'now' situation, but on applications feasible after at least 10 to 15 years from now, for example by assuming drastic miniaturizing of the sensors. CONASENSE is oriented on the mid-term to long-term future. Even when there are no technological limitations cooperation and interference policies have still to be taken into account. All devices should be intelligent so that a future, adaptive, intelligent and green CONASENSE system can be manufactured. Visionary applications should be emphasized. A very limited number of topics (2 or 3) should be selected for a period of for example 2 years. After 2 years new topics can be chosen. The following items have been discussed

(a) Each topic may contain a combination of technological disciplines (e.g., integration of systems, their interferences, software defined radio, cognitive radio, positioning and navigation, nearby and remote sensing, etc.) which allows for realization of selected visionary applications.

(b) A focus, which received much support from the participants, concerned the topic "health". Within this topic many issues can be raised, e.g., care for elderly people in environments which can vary largely, care for pregnant women, medical care at hospitals, etc. The medical world will be involved in CONASENSE to define thoroughly the wish list in the health community. Cross-connections with a series of disciplines should be emphasized.

(c) Other CONASENSE topics mentioned are traffic management systems, security, scanning and space-based services. Definitive choices will be made in 2013.

1.3.1.3   Parties in CONASENSE

In CONASENSE scientific organizations and organizations for developing and producing test-beds and demonstrator systems as well as organizations interested in using CONASENSE proto-type systems in specific applications should be amply represented in the Foundation. Each organization is represented by becoming a member of CONASENSE. Members can come from organizations with CONASENSE specialists and/or CONASENSE decision makers, from Academies of Science in various countries and from EU and UN related organizations and political organizations. A membership data base needs to be set up in the present initial phase of CONASENSE.

The broad areas of potential CONASENSE applications in various sectors of the Society necessitate that we should organize open brainstorm sessions per sector. The presenters of the session are internationally-recognized key players within the sector in academia, research institutes, industries and most innovative user organizations. The expected impact of applications in various sectors requires new procedures for CONASENSE knowledge protection and/or transfer approved by the members. The Foundation is open for all countries. Attention will be paid that big countries from all continents are well represented. In many countries national projects are underway or will be proposed and often allow for international cooperation.

CONASENSE topics should be interesting for BSc and MSc students and PhD's. Young bright students and PhD's can be attracted to do research on a CONASENSE topic and invited for coming to the CONASENSE workshops. Student exchange is at the moment very common, especially within Europe.

EU, but also ASIAN, programs in for CONASENSE interesting areas are huge. A strong CONASENSE consortium opens better opportunities for granting an EU project. Working groups can do the preparatory phase. Full scale activities should be executed by selected partners in CONASENSE. Also KNAW-like organizations should be approached to check if those organizations are willing support official recognition.

1.3.2 CONASENSE Futuristic Workshop Per Topic

As said in the first part of this chapter we pay in the initial phase of CONASENSE large attention to organize brain-storming workshops in which experts and specialists from universities, institutes and industry with knowledge in one (or max two) areas of CONASENSE can bring in their points of view.

The so-called futuristic workshop has two aims: establishing the CONASENSE topical roadmap and selecting bright/young MSc graduates and post-docs from anywhere in the world, who are interested in doing basic/fundamental and applied research for short/medium/long-term developments on the topic. From the roadmap we derive a short-term plan

(1–5 years from now), a medium-term plan (5–20 years from now) and a long-term plan (20–50 years from now).

The futuristic workshop sequence is in the start-up phase once per year, preferably March, April because in those months no other prime (related) conferences are being held. Duration is 2 to 3 days; place in the start-up phase will be Amsterdam (good for building-up identity). Top experts in a selected (multi-disciplinary) CONASENSE topic are invited to give an introductory overview on the importance of their discipline in connection to the topic and to stimulate the brainstorming process. From all presentations and discussions the experts draft the roadmap. They are also invited to write a long article in which they present their long-term visions and derived from the visions the short- and medium term plans. The selected topics will get ample attention in two workshops covering a period of two years. After these two years the progress on the topics of two workshops will be reported in a new CONASENSE Journal and technical books while new topics will be chosen for coming futuristic workshops prepared for the next period of two years.

Futuristic Workshop publishing is done by River Publisher. Per year 2 to 3 journals issues and several books will be published with workshop presentations and/or long articles of invited workshop speakers.

After the start-up phase we plan to organize an extra day immediately after each workshop. On this day a symposium and forum will be held in which key decision makers can give feedback to outcomes of the workshop with respect to visions and roadmap. Furthermore, they can indicate how this feedback can be used to concretize steps leading toward major future policies and strategies for optimizing opportunities and chances per topic or theme to set up CONASENSE test beds and to develop CONASENSE demonstrator systems.

Because brain-storming is a crucial aspect in the workshop the number of attendees will be limited to a maximum of one hundred. A workshop fee will be raised to cover all organizing costs.

1.3.3 CONASENSE Organization

The organizational structure of CONASENSE Foundation consists of a Board with a chairman, secretary and board members, an administration, a review board, working groups and a liaison officer.

The Board is responsible for the vision, objectives and the short, medium and long term roadmaps.

Major tasks of the review board are related to reviewing the journals, books and the outcomes of working groups. Working Groups consist of specialists interested in specific areas of research which comes under the umbrella of CONASENSE.

The liaison officer is responsible for liaising with multi-national organizations including organizations under the EU, UN, ITU and standardisation bodies. Annual membership fees will be moderate, lowest for students and highest for large non-academic organizations and companies.

Dissemination from CONASENSE activities will take place via the CONASENSE Magazine once per year, CONASENSE Journal with 2 to 3 issues per year and technical books. The publisher for these forms of dissemination will be River Publishers. At present we are still working on the working structure for the magazine, journal volumes and books.

1.4 Conclusion

CONASENSE has potentials in a wide variety of societal areas and therefore the topics can be very diverse. However, I think the choice for the topic health in the initial phase of CONASENSE for the coming 2 years is already challenging enough. After 2 years new topics can be chosen.

Problems addressed per topic that can not be solved with existing knowledge in near future, but may be solved in coming 20 to 50 years from now based on breakthroughs in science, technology and systems and on gained insights in utilization potentials: that is what CONASENSE is about.

References

[1] Lian, X., H. Nikookar, and L. Ligthart, "Application of Green Radio to Maritime Coastal/Lake Communications and Locationing Introducing Intelligent WiMAX (I-WiMAX)", in *Towards Green ICT*, Chapter 20, Riverpublishers, 2010.

[2] Che, H., "Adaptive OFDM and CDMA Algorithms for SISO and MIMO Channels", PhD Thesis, Delft University of Technology, ISBN 90-76928-08-8, 2005.

[3] Harmer, D., M. Russell, E. Frazer, T. Bauge, S. Ingram, N. Schmidt, B. Kull, A. Yarovoy, A. Nezirovic, L. Xia, V. Dizdarevic, and K. Witrisal, "EUROPCOM: Emergency Ultrawideband RadiO for Positioning and COMmunications", invited, Proc. *IEEE International Conference on Ultra-Wideband*, vol. 3, pp. 85–88, 2008.

Appendix

List of participants

Prof. Dr. Leo Ligthart, Delft University of Technology, The Netherlands
Prof. Dr. Ramjee Prasad, CTIF, Aalborg, Denmark
Prof. Dr. M Rugieri, IEEE-AESS, University of Rome Tor Vergata, Italy
Dr. Homayoun Nikookar, Delft University of Technology, The Netherlands
Prof. Silvano Pupolin, University of Padua, Italy
Prof. Dr. Vladimir Poulkov, Technical Unverisity of Sofia, Bulgaria
Dr. Tomasso Rossi, University of Rome Tor Vergata, Italy
Prof. Dr. Enrico Del Re, University of Firenze, Italy
Prof. Dr. Kwang-Cheng Chen, National Taiwan University, Taiwan
Dr. Ole Lauridsen, Terma A/S, Denmark
Prof. Dr. Erik Fledderus, TNO-ICT, The Netherlands
Prof. Dr. Mehmet Safak, University Hacettepe, Turkey
Dr. Silvester Heijnen, CHL, The Netherlands
Prof. Dr. Durk van Willigen, Gauss Research Foundation, The Netherlands
Rajeev Prasad. River Publishers, Denmark

Biography

Prof. dr. ir. Leo P. Ligthart, Ceng, FIET, FIEEE was born in Rotterdam, the Netherlands, on September 15, 1946. He received an Engineer's degree (cum laude) and a Doctor of Technology degree from Delft University of Technology in 1969 and 1985, respectively.

He is Fellow of IET and IEEE.

He received Doctorates (honoris causa) at Moscow State Technical University of Civil Aviation in 1999, Tomsk State University of Control Systems and Radio Electronics in 2001 and the Military Technical Academy Bucharest in 2010. He is academician of the Russian Academy of Transport.

In 1988 he was appointed as professor on Radar Positioning and Navigation and since 1992, he has held the chair of Microwave Transmission, Radar and Remote Sensing in the Department of Electrical Engineering, Mathematics and Computer Science, Delft University of Technology. In 1994, he founded the International Research Center for Telecommunications and Radar (IRCTR) and was the director of IRCTR until 2011. He received several awards from Veder, IET, IEEE, EuMA and others.

He is emeritus professor at the Delft University of Technology, is guest professor at ITB, Bandung and Universitas Indonesia in Jakarta and scientific advisor of IRCTR-Indonesia. He is founder and chairman of Conasense. He is member in the Board of Governors IEEE-AESS (2013–2015).

He is founding member of the EuMA, organized the first EuMW in 1998, the first EuRAD conference in 2004 and various conferences and symposia. He gave post-graduate courses on antennas, propagation and radio and radar applications. He was advisor in several scientific councils and consultant for companies.

Prof. Ligthart's principal areas of specialization include antennas and propagation, radar and remote sensing, but he has also been active in satellite, mobile and radio communications. He has published over 600 papers and 2 books.

2

Integration of Communications, Navigation, Sensing and Services for Quality of Life: Challenges, Design and Perspectives

Marina Ruggieri[1], Ramjee Prasad[2],
Mauro De Sanctis[1] and Tommaso Rossi[1]

[1]*CTIF Italy (University of Rome "Tor Vergata"), Italy*
[2]*CTIF HQ (Aalborg University), Denmark*

2.1 The "Integrated Vision"

All possible visions of future services for the information society have to start with the identification of user requirements. In this context the service shall be able to improve the users Quality of Life (QoL). To this respect, user satisfaction is expected to be gained by augmenting the set of services that can be supported and by improving the performance of each service in an easy to use, efficient and secure fashion. On one hand, a method for improving the performance of a system is to increase its components/entities thus enhancing the number of design solutions. On the other hand, a higher number of system components/entities also extends the set of available services.

A very important integration strategy concerns communications, positioning and sensing systems. This integrated vision involves an "active" integration with new business opportunities able to merge three worlds — communications, navigation and local/remote sensing — that have been apart for years. This vision is the focus of the Communications, Navigation, Sensing and

COmmunications- NAvigation-SENsing-SErvices (CONASENSE), 19–41.

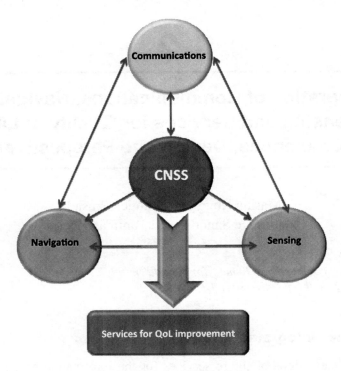

Fig. 2.1   Integrated CNSS for QoL.

Services (CNSS) paradigm for QoL services provision. In CNSS, communications, navigation and sensing systems can mutually assist each other by exploiting a bidirectional interaction among themselves (Figure 2.1).

Next generation communication systems will require the support of a variety of services and applications [1]. The subscriber will demand services at different locations having a seamless single subscriber device. This evolution requires an effective integration of communication networks as well as location-dependent services and context sensing for providing the service and evaluating the current situation. Therefore there is a need for a common platform aimed to the synergistic combination of technologies for communications, positioning and sensing systems able to effectively use information about location, context and situation of several entities.

The term "context", often overlaps with "location" and "situation". In more detail, the term location refers to the geographical coordinates of mobile users and in some cases it can be extended by including other information such as speed, acceleration, direction and orientation. As an example, a network control centre can process the information about the location of the mobile user with the aim at computing: user location with respect to the cell of coverage; user distance from the access nodes; path and next location of the user/node. In satellite networks, high altitude platform networks and ad-hoc networks, network nodes are mobile in nature. Therefore, in such networks, the term location can be also referred to the geographical coordinates of the network nodes. Through this knowledge, the integrated system can manage the dynamics of the network topology, the coverage and also the resource distribution more efficiently.

The term context is referred to a set of parameters that can be used to describe the environment in which the user is embedded and the devices and access networks with which the user interacts. Several categorizations have been proposed to structure context information. Often they are oriented on a particular application domain. An example of generic categorisations is to divide the context into: system context, user context and environmental context. System context represents the information related to both the computing system it is running on (e.g., the particular type of mobile device) and to the communication system being used (e.g., the particular type of wireless network). System context deals with any kind of context information related to a computing system, e.g., computer CPU, access network, network address, status of a workflow, etc. User context can be quite rich, consisting of attributes such as physical location, physiological state (e.g., body temperature, heart rate, blood pressure, respiration rate, etc.), emotional state, activity state (e.g., talking, walking, running, etc.), and daily behavioural patterns. Environmental context includes lighting, sound, humidity, air quality, temperature, weather, etc. In this frame, sensors play a key-role in several applications. Examples are provided by medical sensors in health monitoring, photonic sensors for robots in the area of universe observation, inertial optoelectronic sensors for Earth observation, photonic nanosensors for scientific payloads, optoelectronic sensors for stress and vibration detection in transport systems, inertial nanoelectronic sensors for satellite navigation, etc.

Parameters that characterize the environmental context are: environmental temperature, environmental noise level, environmental light level, etc. Sensor devices and local/remote sensing systems could be used to get environmental contextual information.

The location information can be considered as part of the user context, but it is more appropriate to split the geographical user location (geographical coordinates x, y, z) which has been previously defined, from the environmental user location (e.g., stadium, hospital, city traffic, train, airplane, etc.). Note that, in certain cases (e.g., in a train, in an ambulance, etc.) the information about the geographical user location is not useful to determine the environmental user location.

The term situation refers to the interpretation of the physical, social or environmental contextual information that can be related to a user and/or an access network. It is worth noting that, whilst context is an objective description of the environment, the situation is a subjective interpretation of a context; this interpretation requires a set of rules defined/personalized by the user by means of the user profile. The user profile defines the mapping rules between context and situation; in particular it defines the way a user wishes to exploit a given resource and, hence, can be seen as a personalized description of the action that the user would like to be performed in a given contextual state. Furthermore, the user profile can be used to provide some personal information such as gender, age, knowledge, emergency state, type of user (e.g., business, traveller, private), type of terminal and its status (e.g., group membership, media supported, reconfigurability, computational capability, battery level/energy resources), money availability and some preferences such as screen colour, font type, level of quality, level of security, the maximum cost that the user is willing to pay for that service, etc. During the exploitation of a service, the user is allowed to change his/her user profile which can be stored into one of the user devices (e.g., a smart card).

Location/context-awareness refers to the ability to use location/context information for different objectives. A system is location/context-aware if it can extract, interpret and use location/context information and adapt its functionality to the current situation of use.

CNSS will address the issues of a smooth evolution toward next generation infrastructure of communication systems, involving heterogeneous networks with location/context awareness. The combined CNSS system will provide enhanced and personalised services, whereas the location/context information will be extremely significant for achieving the required quality to added value services. The focus of CNSS is on delivering enhanced services for the right person, at the proper place, in due time, with the required quality for a large variety of applications including emergency and healthcare assistance. "Enhanced Service" means to provide context/situation-based services; "Right Person" means to allow user personalisation, security and privacy; "Proper Place" means to provide location-aware services; "Due Time" means to provide services in a timely manner; "Required Quality" means to improve service quality by means of radio resource management.

2.2 Benefits for Quality of Life Improvement

The goal of future integrated services will be the improvement of users QoL. The integration concept previously described is the key for the achievement of this objective. In this Section the benefits of integration of communication, navigation and sensing are briefly summarized; in the following sections some practical integration strategies and services are described.

2.2.1 Communications Systems

Radio Resource Management Using Position and Context Information

Localization technology has reached today a good level of accuracy and resolution. This has recently led to the strong interest toward location-based services. However, once that this information has been made available to the user and/or the network, it could be used for other purposes than providing services to the user. Some works have already shown that these information can be used to improve radio resource management or mobility management (i.e., horizontal handover) by properly designed mechanisms. We claim that location information, together with the knowledge of the user situation or context, could become the most important enabling function for design solutions over heterogeneous wireless networks, in order to provide efficient integration of different access technologies. The exploitation of location and context information for managing radio resources of heterogeneous access networks can improve the quality of service and availability.

Context/Location Based Communications Services

There are several examples of location based services integrated with communication systems that can improve the attention to communication services: location-based advertising, location-based alert notification, friend finder, search and rescue. Context information such as gender, culture, time, network congestion etc. can be integrated with communication systems to provide context-based services and to improve the network utilization on the basis of the user preferences and network status.

2.2.2   Navigation Systems

High Precision Navigation

The performance of satellite navigation systems if affected by the non-uniform velocity of electromagnetic waves through the Earth atmosphere. Remote sensing systems can be used to extract useful information regarding the ionospheric conditions so that correction can be applied to the solution of navigation equations thus improving navigation data precision.

Indoor Navigation

Most of the current navigation systems are based on global satellite constellation signals and this type of signals can not be received properly in indoor conditions or within urban canyons. In these cases, the location information can be provided partially or totally by a wireless communication system used as positioning system.

Assisted Satellite Positioning

In bad signal propagation conditions, the startup time of a satellite-based positioning system can be very high. However, the startup time can be lowered by using a wireless communication system combined with a processing system which can be used to communicate the ephemeris of the satellites and/or to compute the position.

2.2.3   Sensing Systems

Delivery of Sensed Data Through Communication Systems

Once that the sensed data has been acquired by the sensing system, these data must be delivered through a communication system to the final user for the proper utilization. Depending on the nature of the sensed data, the requirements on the communications system can range from high data rate delivery to energy efficient low data rate delivery and from real time to near real time.

Integration with Positioning Systems

The information contained in the sensed data can be enhanced by including the georeferentiation thus integrating context with location. While this is a common strategy in satellite-based remote sensing.

2.3 Design of Integrated Systems for QOL: Integration Strategies

Current system designers need a full awareness about the deep changes that the design of complex systems and services toward "beyond 4G" implies. As a matter of fact, the convergence of technologies pushes toward the effective exploitation of all components (terrestrial, air and space-based) and media (wired, wireless) in a fully integrated and possibly seamless way to the end-users. "Integration" (I-concept) becomes, hence, the key-word for the achievement of this complex goal, at coverage, system, technology, service and user terminal levels.

It is clear that the carrying out of I-concept needs to face some challenges, at design and implementation level [2, 3].

Design of integrated communication networks

One of the main challenges to effectively put in practice the I-concept is to identify the "glue" of the integrated system, the communication network architectures.

In the "beyond 4G" communication perspective future architectures may include various global components: cellular communications, space- and air-based communications, wireless local and personal/broadband area communications. The overall network structure generates a variety of services and system features by the flexible and/or dynamic exchanges among the various layers and within each layers. Each layer envisages the use of different frequency ranges, technologies and hardware integration level. The above model, that depicts the technological coexistence and the necessary interfaces, certainly provides hints on the complexity of a global design approach and the related tools. The global model exploits mainly terrestrial facilities in terms of network nodes and connections. However, the satellite and aerial components can be usefully considered to complement coverage, provide a back-up to the terrestrial section, provide additional services, strengthen the services mostly provided by the terrestrial facilities. In the difficult aero-space environment, the development of the I-concept passes through a wise exploitation of four basic ingredients: a flexible *design* at sub-system level, a suitable *technology* [4, 5], a *networking*-oriented approach in the design of system architecture and a fearless approach in facing the various *challenges* that can be met in the deployment of the system.

The vision where the aero-space section of the above model is conceived and developed in full harmony with the other components-aiming at an effective integration for the benefit of users and operators — has been named I-concept, where "I" stands for "Integration" [2]. The integrated vision translates in the layered architecture depicted in Figure 2.2. As it can be noticed, the aero-space component is itself structured in various layers: the HAP (high altitude platform) layer, where manned or unmanned stratospheric vehicles are located at about 20 km height, and the layers where satellites can be mainly located (low, medium and geostationary orbit; LEO, MEO and GEO in acronyms, respectively, as well as highly elliptical — HEO-orbits, not displayed in the figure).

Due to the non uniform nature of the various components in Figure 2.2, the layered architecture needs various modeling and simulation approaches

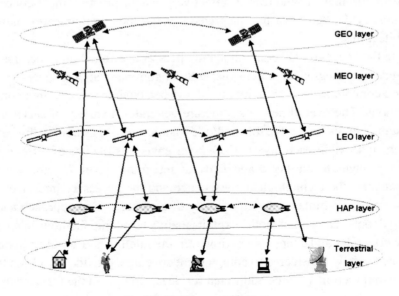

Fig. 2.2 Multi-Layer architecture.

that then have to be properly interfaced for a proper exploitation of the global system.

Examples I-Systems Design

The satellite navigation and the communications world share various commonalities and, at the same time, complement each other. This concept is particularly interesting, accounting for the dramatic penetration of navigation services in people's everyday life and in the security tissue of nations [6] as well as the key-role of mobility in the communications world. A synergic cooperation between navigation and communications services can overcame the receive-only nature of the navigation services, adding interactive communications capabilities, and enhancing the communications service through the navigation-related data.

A new class of users — the *NavCom* users — will continue to grow in the next years. The new type of services and users translate into an extension of the connectivity layer that has to link and interface not only wireless and wired communications networks but also satellite navigation ones; furthermore, also the service-awareness concept has to be modified to replace the basic communications service with the NavCom service.

It is worth mentioning that wireless and wired "communication networks" that the authors refer to, include the "sensors networks" that are able to provide additional, and sometimes "vital", data/information to for the service supply.

The large variety of services that are planned or potential for GNSS-2G (Global Navigation Satellite System-Second Generation) is already pushing the navigation user to the need for communications capabilities, hence to become a NavCom user. The GNSS-2G architecture is prone to the NavCom integration, either at global or at local component levels.

Therefore, NavCom users will be found in each layer of the GNSS-2G model [6]:

- *land/water layer* (Earth surface, oceans/seas/lakes/rivers)
- *air layer* (atmospheric region)
- *space layer* (above atmospheric region to deep space).

An interesting scenario for NavCom integration concerns transportation systems and, in particular, trains, as depicted in Figure 2.3.

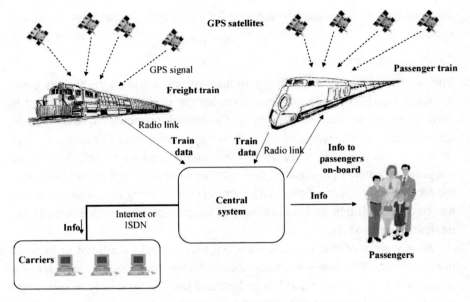

Fig. 2.3   GNSS-based passenger and freight transportation.

In Figure 2.3 a possible integrated architecture is shown, where a satellite-based central system is able to inform passengers about the next station, the elapsed time before the next stop as well as the eventual late arrival, by means of TV screens or LCD monitors. Additional information about sights along the track can be given by using interactive monitors. The service is provided through an integration between GNSS and terrestrial communications links.

Another interesting application of NavCom to transportation system is the car assitance in highway, parking and limited traffic zone tolling.

Further NavCom architectures derive from the integration between the GNSS and existing or planned satellite communications networks, like INMARSAT, IRIDIUM, GLOBALSTAR, THURAYA and QZSS [6].

The challenging deployment of global integrated networks can take large benefit from the exploitation of the stratospheric component of the layered model depicted in Figure 2.2. The reduced time-to-deployment, that is intrinsically available by using HAPs, can represent an effective aid in the graceful development of complex integrated architectures [7] and for emergency applications that require a fast response to events that cannot be predictable [8].

In the above frame, the major exploitation areas of HAP in an *I-Net* can be referred as:

- *permanent* building block (i.e., with the same time duration of the whole network)
- *on-demand* block that can be divided into:
  - *temporary* block
  - *emergency* block

As *temporary* block, the HAP can be needed for a graceful deployment of a complex network, whose core functions have to be rendered operational in a short time, with respect to the natural time expected for the whole network development.

The temporary block can be usefully exploited also for a traffic emulation in pre-operational conditions. A further application is the technological validation of specific sub-systems or devices.

As *emergency* block, the HAP can be exploited both to overcame a happened disaster or to prevent its consequences. It can usefully bypass stations of either terrestrial or space networks that have been damaged either by natural events or intentionally (e.g., terrorist attach scene). In a prevention mode, the HAP can be deployed to create a back-up to the network segments in danger.

2.4 Services for the Short/Medium/Long-Term

As a consequence of the previous considerations, a multitude of services for the information society, having the goal of QoL improvement, can get benefit from the integration of communications, navigation and sensing systems. Some examples are in the area of: Air Traffic Management (ATM), real-time alert systems, nowcasting, Earth science and interplanetary space science, disaster monitoring, safety critical services, e-Health, etc.

In the following, a (not-exhaustive) list of medium term services is proposed.

Disaster Monitoring and Real Time Management

In case of large scale disasters such as floods, earthquakes, forest fires, processed Earth Observation data can be used to predict in real time the evolution of the situation. This processed geo-referenced data can be communicated

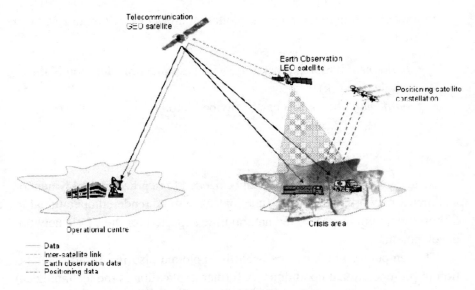

Fig. 2.4   I-System for disaster monitoring and real time management.

to the first responders so that disaster management resources can be pre-positioned to high risk areas. An example of integrated architecture for disaster monitoring and real time management is depicted in Figure 2.4.

Innovative architectures and technologies for earthquake disaster ahead management (DAM), based on the "integrated vision" described in this Chapter, are reported in [9].

Telemedicine services for e-health

The so-called e-Health is currently a very important topic of research; including a large set of medicine/healthcare services provided through ICT technologies. The nature of these services include the integration of sensing and positioning data that are exchanged in a very heterogeneous network structure [10]. This kind of service requires, by its nature, the application of the I-concept previously introduced; an example is the integration of "local" sensor/positioning networks able to monitor some parameter of a patient, in a wider network for remote supervision by a medical center.

Mountain Rescue

Search and rescue operations in mountain areas have particular characteristics. First, there are multiple search and rescue teams in a possibly large area that must be coordinated. Second, there are unsearched areas that are discovered only a posteriori. Therefore, it is needed to coordinate the operations between the teams and the operation management center through a communication system with a wide area coverage. Furthermore, the position of each rescuers should be known and shared with the other rescuers and the history of searched/unsearched locations should be available on a high resolution real-time update map collected from a remote sensing satellite system.

Maritime Surveillance

Maritime surveillance is carried out by national authorities mainly to identify and deter infringements to regulations and security and safety threats. In this scenario the integration of Communications, Navigation and Earth Observation systems is of paramount importance. Earth observation data can be used to detect infringements to regulations, while navigation data must be used to reach the target. In this case, a virtual crisis room between coastal protection bodies and other governmental bodies through video conference is an important tool.

Furthermore, updated meteorological data is a useful information for the ship surveillance especially in case of bad weather conditions which can determine the result of a mission.

Road Traffic Optimization

A possible integrated CNSS service can employ a EO system able to provide images and information related to road conditions, in terms of traffic or road accident in real time. These data can be integrated with those, coming from a local network or system (e.g., cameras), and a localization system network. Finally the TLC system transfer the elaborated information to every single user provided with a NAV system making him able to optimize the route, and reducing, at the same time the traffic congestion.

Furthermore, Electronic Toll Collection (ETC) should use effectively a CNSS system for a new technological implementation of a road pricing concept.

Oil Spill

Integration of satellite sensors can be of great support for the O&G industry. As an example, an Oil & Gas Company which operates a very long pipeline network stretching for hundreds for miles across a very remote and not easy to access area, needs near real time updates on the status of its pipeline, to precisely know what is happening, and where, in order to detect and prevent possible leaks that might be caused by any breach in the pipe works during normal operations. In this case, a remote monitoring system is crucial to prevent or fast intervene in case of leakage from the pipeline.

Precision Agriculture/Farming

Precision agriculture (or precision farming) is the name of an application of positioning, and remote sensing technologies to large farms.

Geographical Information System (GIS) maps are used to determine which fields are moist or dry, and where there is erosion of soil. The processed data can be used by the farmer to automatically regulate the use of fertilizer, pesticide, water through automatic guidance of farm vehicles.

Communications systems are also required for the real-time update of processed maps and eventually for the remote control of farm vehicles.

2.5   Final Remarks

In this chapter a vision of future global networks, as integrated Nav-Com-Sensing networks, for the provision of services improving users QoL is provided. Moving from the integration concept and *I-System* definition, some NavCom architectures have been introduced where the terrestrial, aerial and space components of the network layered model are effectively exploited for the provision of complex QoL services. In this frame, the stratospheric realm has been recognized as a useful means for complex network deployment and service provision.

References

[1]   Simone Frattasi, Rasmus L. Olsen, Mauro De Sanctis, Frank H. P. Fitzek, Ramjee Prasad, "Heterogeneous Services and Architectures for Next-Generation Wireless Networks", *2nd International Symposium on Wireless Communication Systems 2005 (ISWCS 2005)*, Siena (Italy), pp. 213–217, September 5–7, 2005.

[2]   E. Cianca, M. De Sanctis, M. Ruggieri, "Convergence towards 4G: a Novel View of Integration", *Wireless Personal Communications, an International Journal*, Kluwer, vol. 33, no. 3–4, pp. 327–336, June 2005.

[3]   R. Prasad, M. Ruggieri, "Technology Trends in Wireless Communications", Artech House, Boston, 2003.

[4] E. Cianca, T. Rossi, A. Yahalom, Y. Pinhasi, J. Farserotu, C. Sacchi, "EHF for Satellite Communications: The New Broadband Frontier", *Proceedings of the IEEE*, Volume: 99, Issue: 11, Nov. 2011.

[5] C. Sacchi, A. Jamalipour, M. Ruggieri, "Aerospace Communications and Networking in the Next Two Decades: Current Trends and Future Perspectives [Scanning the Issue]", *Proceedings of the IEEE*, Volume: 99, Issue: 11, Nov. 2011.

[6] R. Prasad, M. Ruggieri, "Applied Satellite Navigation Using GPS, GALILEO, and Augmentation Systems", Artech House, Boston, 2005.

[7] M. Ruggieri, E. Cianca, "HAP-Based Integrated Architectures for NavCom", *Workshop on Broadband Access via High-Altitude Platforms*, BBEurope 2005, Bordeaux, December 2005.

[8] G. Araniti, M. De Sanctis, S. C. Spinella, M. Monti, E. Cianca, A. Molinaro, A. Iera, M. Ruggieri, "Cooperative terminals for Incident Area Networks", *First International Conference on Wireless VITAE 2009*, Aalborg (Denmark), pp. 549–553, 17–20 May, 2009.

[9] M. Buscema, M. Ruggieri (Eds.), "Advanced Networks, Algorithms and Modeling for Earthquake Prediction", River Publishers, 2011.

[10] A. Kocian, M. De Sanctis, T. Rossi, M. Ruggieri, E. Del Re, S. Jayousi, L. S. Ronga, R. Suffritti, "Hybrid satellite/terrestrial telemedicine services: Network requirements and architecture", *IEEE Aerospace Conference 2011*, Big Sky (Montana), pp. 1–10, March 4–12, 2011.

Biography

Marina Ruggieri

Academic Positions:

- *Full Professor* in Telecommunications at the University of Roma Tor Vergata (since 2000).
- *Director* of the M.Sc in "Advanced Satellite Communications and Navigation Systems" at the University of Roma Tor Vergata (since 2003). The M.Sc has 40 industrial, governamental and scientific partners; it is CERMET certified.
- *Dean's Deputy*, Service Center of the Engineering Faculty (since 2005).
- *Vice President*, Teaching and Scientific Committee of the University of Tor Vergata Distance Learning School (IAD) (since 2011).
- *Member* of the Strategic Committee of the University of Roma Tor Vergata.
- *Associate Professor* in Telecommunications at University of L'Aquila (1991–1994) and at the University of Roma Tor Vergata (1994–2000);
- *Research and Teaching Assistant* at the University of Roma Tor Vergata (1986–1991);

Industry Position:

- Microwave Integrated Circuit Engineer at FACE-ITT (Pomezia, Italy) and GTC-ITT (Roanoke, VA) in the High Frequency Division (1985–1986).

Governamental Positions

- Member of the Technical-Scientific Committee of the Italian Space Agency (ASI) (2004–2006)
- Vice-President of the ASI Technical-Scientific Committee (2007–2008).

- Member of the Italian *Superior Council of Telecommunications* as Expert (since 2007).
- Member of the Committee of Experts for the Research Policy (CEPR) of the Ministry of University and Research (MIUR) (since 2010).

International and National Positions

- Vice President of the *AFCEA Rome Chapter* (since 2006).
- Italian representative in the Technical Committee on *Communications Systems* (TC6) of the International Federation for Information Processing (IFIP) (2006–2007).
- Member of the Presidential Committee of the Italian Institute of Navigation (IIN) (since 2010).
- Member of the Telespazio Committee for the Innovation Award (since 2006)

IEEE Activities

Division IX Director-Elect (2013)
Division IX Director (2014–2015)
Member, TAB Strategic Planning Committee (2011–2013)
Member, Technical Activities Board (2010–2011)
TAB Representative, Women in Engineering Committee (2011)
President, IEEE Aerospace and Electronic Systems Society (2010–2011)
Executive Vice President, IEEE AESS (2008–2009)
International Director of Italy & Western Europe, IEEE AESS (2005-present)
Member, IEEE AESS Board of Governors (2000–2002)
Member, IEEE AESS Board of Governors (2003–2005)
Member, IEEE AESS Board of Governors (2007–2009)
Chair, IEEE AESS Space Systems Panel (2002–2010)
Vice Chair, IEEE AESS Space Systems Panel (2010-present)
Editor, IEEE Transactions on Aerospace and Electronic Systems (2001-present)
Assistant Editor, IEEE Systems Magazine (2007–2009)

Associate/Sector Editor, IEEE Systems Magazine (2005–2006)
Member, IEEE Judith A. Resnik Award Committee (2002–2003; 2004–2005)
AESS Representative, IEEE Aerospace Conference (2002-present)
Various positions on Committees of IEEE Conferences
Various positions on Committees of IEEE technically-cosponsored Conferences
2011 AESS Service Award
IEEE S'84-M'85-SM'94

Awards and Nominations

- *2013 Roma Capitale Donna* (for scientific research)
- *2009 Pisa Donna* Award (for the career as a women in engineering)
- *2009 Best Paper Award in poster category*, 2nd International Symposium on Applied Sciences on Biomedical and Communication Technologies ISABEL (for the paper *A Novel FM-UWB system for Vital Sign Monitoring and its Comparison with IR-UWB*).
- *2000 Excellent Paper Award*, 3rd International Symposium on Wireless Personal and Multimedia Comunications, Bangkok (for the paper *Adaptive Orthogonal Codes Assignment in CDMA Satellite Systems*).
- *1990 Piero Fanti International Prize* (for a Q-band satellite system)
- Nomination for the 1996 *Harry M.Mimmo Award* (for the paper *A Reliability Model for Active Phased Arrays in Satellite Communications Systems*, Proceed; IEEE Symp. on Phased Array Systems and Technology, Boston).
- Nomination for the 2002 *Cristoforo Colombo Award* (for a W-band satellite system).

Main Research Topics:

- Space communications systems, in particular related to the use of extremely high frequencies, like Q/V and W bands. In particular she is the Principal Investigator of the Communication Experiment of the ESA Alphasat TDP#5 scientific mission, that will deploy in 2013 the first European civil satellite communications payload in the Q-V band (40–75 GHz). The control & data center of the Communication Experiment will be located in the University of Roma Tor Vergata (Italy).

- Integration of terrestrial, satellite and aerial communications systems for civil, military and dual applications.
- Infrastructures, applications and innovative services for the *Information and Communications Technology* (ICT).
- Global Satellite Navigation Systems and related applications.
- Development of integrated navigation-communications (nav-com) prototypes for various applications (e.g. maritime, tourism, high-speed cars, robot set)
- Wireless networks.
- ICT for Quality of Life (QoL) in particular in the following areas: *Health, Energy, Biotechnology, Disaster Ahead Management, Law* and *Cultural Heritage*

Research Center Management

- *Director* of *CTIF_Italy*, the Italian branch of the *Center for Tele-infrastruktur* (CTIF) in Aalborg (Danimarca), opened at the University of Roma Tor Vergata (since 2006).
- A CTIF_Italy research office has been opened in the industrial facility of *Space Engineering/TES* in Tito (PZ, Italy) (2010).
- A CTIF_Italy research office is going to be opened (MoU signed) in the industrial facility of *Integrated Software Systems* (SSI), Taranto, Italy.
- A number of technical and scientific agreements have been signed between CTIF_Italy and industries or scientific institutions.
- A laboratory facility for the development of integrated navigation and communications prototypes has been opened in the CTIF_Italy premise at the University of Roma Tor Vergata (2010). The laboratory is named **HASCON** (**H**ardware and **A**lgorithmic **S**olutions of **CO**mmunications and **N**avigation).
- Patent: De Sanctis M., Monti M., Ruggieri M., Prasad R., "Procedimento per la coesistenza di reti di trasmissione senza fili/Method for the Coexistence of Wireless Networks", in process, ITA02/02/2009.

Publications

- 308 papers (international journals/transactions, proceedings of international conferences, book chapters)
- 12 books.

Ramjee Prasad is currently the Director of the Center for TeleInfrastruktur (CTIF) at Aalborg University, Denmark and Professor, Wireless Information Multimedia Communication Chair.

Ramjee Prasad is the Founding Chairman of the Global ICT Standardisation Forum for India (GISFI: www.gisfi.org) established in 2009. GISFI has the purpose of increasing of the collaboration between European, Indian, Japanese, North-American and other worldwide standardization activities in the area of Information and Communication Technology (ICT) and related application areas. He was the Founding Chairman of the HERMES Partnership — a network of leading independent European research centres established in 1997, of which he is now the Honorary Chair.

He is the founding editor-in-chief of the Springer International Journal on Wireless Personal Communications. He is a member of the editorial board of other renowned international journals including those of River Publishers. Ramjee Prasad is a member of the Steering, Advisory, and Technical Program committees of many renowned annual international conferences including Wireless Personal Multimedia Communications Symposium (WPMC) and Wireless VITAE. He is a Fellow of the Institute of Electrical and Electronic Engineers (IEEE), USA, the Institution of Electronics and Telecommunications Engineers (IETE), India, the Institution of Engineering and Technology (IET), UK, and a member of the Netherlands Electronics and Radio Society (NERG), and the Danish Engineering Society (IDA). He is also a Knight ("Ridder") of the Order of Dannebrog (2010), a distinguishment awarded by the Queen of Denmark.

 **Mauro De Sanctis** received the "Laurea" degree in Telecommunications Engineering in 2002 and the Ph.D. degree in Telecommunications and Microelectronics Engineering in 2006 from the University of Roma "Tor Vergata" (Italy). He was with the Italian Space Agency (ASI) as holder of a two-years research fellowship on the study of Q/V band satellite communication links for a technology demonstration payload, concluded in 2008.

From the end of 2008 he is Assistant Professor at the Department of Electronics Engineering, University of Roma "Tor Vergata" (Italy), teaching "Information and Coding Theory". From January 2004 to December 2005 he has been involved in the MAGNET (My personal Adaptive Global NET) European FP6 integrated project and in the SatNEx European network of excellence. From January 2006 to June 2008 he has been involved in the MAGNET Beyond European FP6 integrated project as scientific responsible of WP3/Task3. He has been involved in research activities for several projects funded by the Italian Space Agency (ASI): DAVID satellite mission (DAta and Video Interactive Distribution) during the year 2003; WAVE satellite mission (W-band Analysis and VErification) during the year 2004; FLORAD (Micro-satellite FLOwer Constellation of millimeter-wave RADiometers for the Earth and space Observation at regional scale) during the year 2008; CRUSOE (CRUising in Space with Out-of- body Experiences) during the years 2011/2012. He has been involved in several Italian Research Programs of Relevant National Interest (PRIN): SALICE (Satellite-Assisted LocalIzation and Communication systems for Emergency services), from October 2008 to September 2010; ICONA (Integration of Communication and Navigation services) from January 2006 to December 2007, SHINES (Satellite and HAP Integrated NEtworks and Services) from January 2003 to December 2004, CABIS (CDMA for Broadband mobile terrestrial-satellite Integrated Systems) from January 2001 to December 2002. He is serving as Associate Editor for the Space Systems area of the IEEE Aerospace and Electronic Systems Magazine. He co-authored more than 60 papers published on journals and conference proceedings.

Tommaso Rossi received his University Degree in Telecommunications in 2002, MSc Degree in "Advanced Communications and Navigation Satellite Systems" in 2004 and PhD in Telecommunications and Microelectronics in 2008 at the University of Rome "Tor Vergata". From April to June 2009 he joined the CTIF-Japan (Center for TeleInFrastruktur), a research center focused on innovative TLC technologies located at the Yokosuka Research Park (Japan). Form 2012 he is an Assistant Professor at the Department of Electronics Engineering, University of Rome "Tor Vergata", teaching "Digital Signal Processing", "Management of Multimedia Information" and "Signals". He is part of the scientific team that will manage the "Alphasat TDP#5" telecommunication experiments; TDP#5 is a joint ASI-ESA mission and is the first European civil satellite communications payload in the Q-V band (40–75 GHz).

He has been involved in many European Space Agency projects: from 2011 he is working on ESA "AIS End-to-End Testbed" project, as responsible for the analysis and design of digital beam-forming techniques. In 2012 he worked on ESA "Miniaturized Low-Weight Space Antenna for AIS VHF Applications" project, as responsible for digital beam-forming software development and support for antenna array design; in 2007–2008 he has been involved in "European Data-Relay Satellite System" project, as responsible for EDRS-user segment visibility analysis and Q/V-bands link budgets analysis; in 2006 he worked on "Flower Constellation Set and Its Possible Applications" research project, as responsible for the design and optimisation of Flower Constellations for communication, navigation and Earth/space observation applications.

He has been a technical member of University of Rome "Tor Vergata" team working on many projects funded by the Italian Space Agency: "FLORAD" project (2007/2008), as responsible for design, optimisation and analysis of a Flower Constellation of millimeter-wave radiometers for atmospheric observation; "WAVE" project (2004/2007), feasibility study for W-band satellite telecommunication payloads, as responsible for small-LEO mission definition and payload design activities; "TRANSPONDERS" project (2004/2008), feasibility study for Q/V-band satellite telecommunication payloads. In 2005 he worked on Regione Lazio "PISTA" project for the development of low-cost inertial/GPS integrated navigation system, as responsible for the design of a data-fusion Kalman filter.

His research activity is focused on Space Systems, EHF (Extremely High Frequency) Satellite and Terrestrial Communication, Satellite and Inertial Navigation Systems, Digital Signal Processing.

He is author of more than 60 papers published on journals and conference proceedings.

3

Flexible Intelligent Heterogeneous Systems for Enhancing Quality of Life

Enrico Del Re and Simone Morosi

Department of Information Engineering, University of Florence, Italy

3.1 Introduction

Looking-ahead to the 10–15 years mid-term, users (people and other entities) will require the access, as simple and natural as possible, to a multiplicity of services and available applications (ideally 'an universal service') provided and supported by a convergent and integrated system of technologies ('an integrated system of systems'). To this end ICT can provide a breakthrough by the integration of communication, localization and sensing functionalities made available to clouds of users by a self-consistent heterogeneous and flexible system.

Key features of this strategy are:

— the applicability to an as large as possible set of future services, having diversified and even divergent requirements
— system intelligence (sensing, learning, decision, action), reconfigurability, adaptability, energy efficiency and security
— context awareness (physical, environmental, situational localization)

COmmunications- NAvigation-SENsing-SErvices (CONASENSE), 43–65.

— exploitation of the benefits of clouds of users
— highly efficient communication systems to deliver all the necessary information in due time, where needed, with required quality of service (QoS) and energy efficiency
— effective and synergistic cooperation among all the potentially available technologies.

Such features identify a system in line with the objective which have been explicitly defined by the European Initiative Horizon 2020, with a clear remind to the "Better Society" concept (i.e., "Longer and Healthier Lives" and "Inclusive Innovation and Secure Society") [1].

All these key issues can be addressed by defining a reconfigurable heterogeneous telecommunication infrastructure (a platform) which can integrate localization and sensing capabilities, terrestrial and satellite communication subsystems and intelligent objects and networks with the ultimate goal of improving the quality of life of the citizens and users in several common scenarios, such as for example:

— people experiencing daily living difficulties at home (e.g., elderly people, chronic disease patients)
— people faced with a sudden emergency situation
— searching a Missing Person
— better user mobility in urban environment
— emergency and crisis management.

These innovative platforms must be highly reconfigurable in order to adapt to different application contexts and exploit the most recent intelligent distributed computing and social interaction technologies (i.e., software agents and cloud computing) by suitable middleware implementations: the objects establish autonomously social relationships of different kinds (e.g.: co-work, co-localization, co-ownership, transactional) and end up composing, maintaining and providing services and information to/from external apps and services. As an example, the system should be able to update the configuration parameter of a handheld device (e.g., an Android Smartphone) to provide the right information when and where needed (e.g., about a person entering a first aid, by providing the right information, such as "your relative is hospitalized in the orthopedic department").

This Chapter is organized as follows. Section 3.2 is devoted to the description of the applications which can improve the quality of life: a particular emphasis is given to tele-health and emergency services. Section 3.3 describes the flexible heterogeneous architectures which can be used to enhance services forthe quality of life, whereas Section 3.4 presents some results which have been already obtained in these fields. Finally, Section 3.5 provides conclusions and some future perspectives.

3.2 Applications for Quality of Life

Due to recent technological and societal changes it is clear that our society is moving towards a future in which telecommunication platforms will be integrated with heterogeneous systems of localization and sensing capabilities and in which intelligent objects will cooperate as in human social networks with the goal of providing services to the people at different level of complexity. To achieve such a visionary scenario, each component has to be designed (and often redesigned) taking into account a remarkable level of generality and universality that characterizes the new telecommunication entities and the new architectures and infrastructures have to be defined considering a plurality of services and applications with a low (ideally with no) reset effort. This platform can support applications in agreement to the objective defined by Horizon 2020.

3.2.1 Integrated Localization/Communications/Sensing (LOC/COM/SENS) Networks for Emergency Applications

The typical emergency scenario refers to a situation of public emergency (e.g., fire, earthquake, flood, explosion, big accident), where several rescuers, organized in teams, convene to the emergency area from different locations, possibly with different transportation means, likely equipped with tools for the intervention (e.g., water or stretchers). They belong to one or more emergency response organizations, e.g., police, fire service, and emergency medical services. A general emergency scenario and typical requirements for telecommunications systems are as follows.

3.2.1.1 Emergency scenario

Every member of the rescue team of First Responders (FRs), as well as the Emergency Vehicles (EVs), is normally equipped with a portable radio transceiver, with advanced and integrated localization/communication (LOC/COM) capabilities. Localization capabilities are necessary to determine the terminal position (desired accuracy would be 1–2 m), using both Global Navigation Satellite System (GNSS) services and, in case of lack of a satellite radio link, terrestrial network-based positioning methods. In some cases, also the members of the same team or of other teams may need to know their reciprocal positions, for safety reasons.

The following elements could also be part of an emergency network:

— a communication facility which guarantees the connection (possibly through a satellite/High-Altitude-Platform-HAP) between the Emergency Control Center (ECC) and people operating in the emergency area. It is a temporary, mobile, or transportable station placed at the boundary of the emergency area and acting as a master node for the local network

— the HAP, moved on-demand above the emergency area to provide ad hoc and temporary communications capabilities. It can connect the Mobile Master Node (MMN) with the ECC and possibly the emergency vehicles (which have higher available power than the first responders' terminals) with the MMN

— the satellite, able to guarantee very long distance communications between the MMN and the ECC and the ECC and the emergency vehicle (EV).

3.2.1.2   Requirements for emergency systems

The requirements of radio communication systems for emergency rescue applications which have been issued by international standardization committees have been clearly identified and thoroughly described in [2]. Communication requirements for emergency services shall guarantee that the required information is available to the correct person or organization at the appropriate time. In essence, communications must be timely, relevant, accurate and secure for all actions that may be undertaken: particularly, the efficiency of the emergency operations is dependent upon the ability of the communication networks to deliver timely information among several authorized emergency teams.

The main requirements for providing efficient and effective emergency services can be summarized as follows:

— fast call setup: typical requirements for voice call establishment times are in the range 0.3–1 s, with 0.5 s often cited as the requirement for wide area operation. For voice-over-satellite connections to a remote ECC, such requirement shall be relaxed and call establishment times in the order of 1–2 s are customary

— instant access: the need for radio capacity is increasing during major incidents and accidents, so efforts have to be made to ensure, as much as possible, that adequate communication facilities be available

— quality of service: voice (or data transmission) quality should be adequate to guarantee the understanding (or correct reception) of the message

— seamless radio coverage: possibly throughout the whole served area availability of radio coverage should be guaranteed also under exceptional conditions (e.g., power supply outages, etc.)

— controlled network access: in order to guarantee controlled load of the network, under certain circumstances, priorities or restrictions should be assigned to specific users

— specific functionalities: advanced features, such as group communications and dispatching, security and encryption, and dynamic resource management, would be very useful.

The fulfillment of the above requirements normally remains under the responsibility of a variety of public authorities, such as national ministries responsible for emergency and security, international agencies, European and national police, civil protection agencies, coast guard, fire brigades, local authorities, etc.

3.2.1.3 An integrated approach for emergency situations

In emergency conditions, especially during the intervention and mitigation phases, the availability of LOC/COM/SENS devices lead to remarkable gains in terms of speed of response, completeness and effectiveness of intervention [3, 4]. Because of the different networks in the considered scenario, these benefits can be completely exploited using reconfigurable and interoperable systems.

Efficient and complete emergency systems can be provided by the synergistic use of communication, positioning and global monitoring services: this trend is confirmed by the initiatives of security and safety agency, research institutions, standardization organizations and by the European Commission guidelines.

Therefore, an integrated communication, localization and sensing system is of tantamount interest for all of the emergency management phases. Particular attention should be given to the integration among the techniques, conceived as stand-alone and not necessarily oriented to multi-disciplinary applications.

To this aim, the following subjects have to be carefully considered:

— the methods of collection, analysis and integration of pre- and post-event GMES data, the data fusion techniques among the information coming from earth observation systems and the information collected by sensors deployed on site

— the context evaluation through cognitive approaches in order to identify the best transmission strategies, taking into account the scarcity of resources in an emergency context

— the advanced localization techniques which are based on the use of Global Navigation Satellite Systems (GNSS) assisted by available positioning information, which can be derived from both the communication and monitoring network.

In this framework, better performance can be achieved through the adoption of adaptive cooperative approach based on communication among users, communication network, emergency control and coordinator center and HAPs or mini-satellites.

The emergency systems which are based on the described approach are characterized by meshed heterogeneous architectures based on both satellite and terrestrial segments: in particular, the use of survived networks, i.e., networks which are involved in the disaster but still partially operating, and networks deployed after the critical event in the intervention area should be guaranteed. Alerting and control signaling are defined according to opportunistic, cognitive and cooperative paradigm. In the satellite and terrestrial components particular attention are given to the definition of terrestrial infrastructure and communication protocols for the transmission of the localization and sensing information, collected with frequency and reliability suitable for an emergency scenario.

The integration of sensors and terminals (even wearable) with other emergency networks represents another important issue: these devices, commonly seen as part of the object domain, can transmit biometric information, data, images also taking part in data fusion procedures on the involved area and through cooperative interactions among the devices themselves.

In emergency conditions, especially during the intervention and mitigation phases, the availability of LOC/COM/SENS devices lead to remarkable gains in terms of speed of response, completeness and effectiveness of intervention. These benefits can be completely exploited using reconfigurable and interoperable systems because of the different networks in the considered scenario.

3.2.2 Tele-Health Services

In the last few years the role of ICT (Information and Communication Technology) in the health system has been one of the main investigated research area [5, 6, 7]. In this context the design of an integrated interactive system for telemedicine services is required both for clinical studies and for self-care purposes: indeed the integration of distributed medical competence and clinical information contributes to the quality of medical care. Although prevention is in all Governments Agenda as the strategy to intervene on population life style and behaviour and to maintain as long as possible people in healthy conditions, nowadays there are very few population management programs that try to intervene on the lifestyle and behaviour modification. Indeed, the acute-disease people are about 20% of the population and represent the cause of the 80% of the healthcare expenditure, while the remaining population is responsible for about 20% of the healthcare expenditure.

In the following subsections the Tele-Health services are described, starting from an accurate analysis of both the state of art of the available services and applications and the user (patient and specialist) requirements. In particular both the Self-Care and the Assisted services are considered.

3.2.2.1 Self-care services

Primary prevention represents a big opportunity for interactive systems, with respect to the traditional information campaign, allowing to guide people in adopting correct and healthy lifestyles, by knowing patients feelings, concerns, habits and relationship with their illness.

The Self-Care Services offer a range of tools, functionalities and services supporting lifestyle improvement and wellness. This goal can be reached by means of tools able to convince the user about the value of this change of behaviour and with the capabilities of supporting him/her through an easy and attractive change process. The user-system interaction level can be decided by the user, who can only give essential information to the system or make some biometrics measurements to assess his health status or to be an active user by adhering to specific care plans.

3.2.2.2 Assisted services

Considering prevention as a big opportunity for Healthcare system in order to reduce the impact of chronic disease on the population asks for a growing demand of technological solutions that can:

- support the increasing number of chronic diseases
- increase ICT integrated systems for easy fruition, exchange and management of clinical data, supporting the medical staff, providing easy access to health data
- reduce the number increase of medical errors and Adverse Drug Effects (ADE).

Two different services can support physicians and healthcare professional activities in this environments, aiming at answering these demands: e-clinical studies and tele-consultation services.

E-clinical studies services make easier and more efficient the work carried out by the experts in e-trials thanks to the continuous monitoring and management of incoming clinical data, which include the use of cryptography for privacy enforcement.

Tele-consultation services enable healthcare professionals to share patients data for early diagnosis and clinical cases discussion, allowing people to access periodical clinical check-up for prevention and avoiding medical institutions congestion.

3.2.2.3 The role of patients/citizens

The Health services should also be "accessed" through mobile phones, smartphone or PDA: in particular this feature could provide the opportunity to:

— access to portals that provide information about the health of citizens
— access to virtual communities and support groups and online support where people can share experiences and information but also specific data content in different EHs (Electronic Health Record)
— access to services for Homecare, even to receive "chronic disease management services" to run the care at home
— access to telemedicine and teleservices that enable collaboration among health professionals.

Therefore, in this context, the creation of personalized health networks, namely of networks that are focused on the concept of PHR (Personal Health Record) as a health personalized issue, is of great importance in the social network dedicated to health and diseases. This is also true in the case where the interface is with the ESF (Electronic Health records) in which a citizen, not just the doctor, inserts the data and personalized information on health. In this framework, the citizen becomes the focal point of the new health ecosystem and collaborates and shares information with not only health professionals but also with other citizens. This model is aimed not only to citizen patients but is open to all those who want to improve relations with their health, quicker and easier interaction with the social and health institutions.

3.3 Flexible Heterogeneous Architecture

The general network architecture to be adopted to provide services for enhanced quality of life is defined and two specific solutions for tele-health and emergency scenarios is presented.

3.3.1 Baseline Architecture

The goal of a synergistic use of communications, localization and sensing services, provided by means of meshed heterogeneous architectures based on satellite and terrestrial segments, can be afforded by the use of the most recent cognitive, cooperative and context- and location-aware technologies, with deterministic and opportunistic radio access: a fully interoperable architecture is enabled by the distributed system intelligence and by suitable application programmable interfaces (API).

The system baseline architecture in Figure 3.1 is as general as possible by assuming that the proposed platform can encompass interfaces to any kind of enabled devices (i.e., fixed, portable and handheld ones) and to suitable control centres and also to both wired and wireless public communication networks and to the most general localization and sensing systems. As a result, a key role in the baseline architecture is taken by the middleware, which is aimed at integrating all the heterogeneous components (coming from the separated domains, such as localization, communication, and sensing) in a common

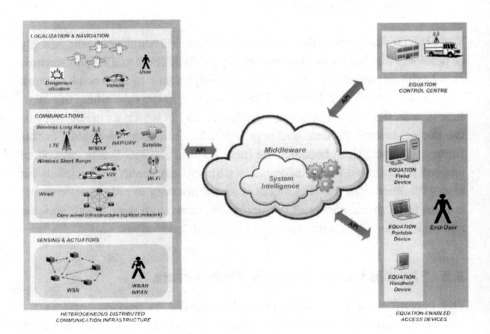

Fig. 3.1 Baseline System Architecture.

stratum that implements the basic functionalities for the discovery, management, composition, and activation of all the services provided by the distributed infrastructure. On top of the middleware the applications are executed, which provide the complex system intelligence functionalities.

The definition of the middleware architecture involves the usage of the newest distributed computing and social interaction technologies, in particular software agents and cloud computing. One issue that needs to be carefully addressed is the definition of an unambiguous description for objects and infrastructure components that "live" in the virtual world defined by Internet of Things (IoT). Another issue is the scalability in discovering and exploiting the services provided by the distributed components.

This architecture also allows the study and the design of advanced and assisted localization techniques based on the use of Global Navigation Satellite Systems (GNSS), like GPS and Galileo, jointly with additional information available by means of a cooperative approach among users. GNSS interference reduction techniques have also to be developed and analysed, taking into account the cooperative/distributed strategies in order to maximize the efficiency of the proposed solutions. The system intelligence supports the interoperability of the sensor networks with the communication systems, so allowing context aware capabilities of the system. In this general framework of reconfigurability, cognitive and cooperative paradigms are adopted, aiming at achieving the benefits which are guaranteed both at the transmission layer and at the network layer: specifically, the cognitive approach is focused on both distributed sensing and the definition of a cross layer solution, where QoS and energy-efficiency requirements are included. The use of the cooperative paradigm can act as a booster to provide general, ubiquitous, energy efficient services for quality of life improvement.

Main drawbacks of this distributed reconfigurable systems can be identified in energy consumption and security/privacy issues. Aiming at increasing the operating time of battery-powered systems, the devices and systems have to pursue the maximization of the energy efficiency with any possible strategies, such as link adaptation techniques, efficient resource management, cognitive/cooperative communications and distribution of heavy computing functionalities by mean of cloud computing.

Both cryptographic techniques to enforce key management procedures and network security protocols, such as IPsec, should be considered to create secure, authenticated, reliable communications over IP networks, providing a high efficiency security procedures to protect IP datagrams (IPv4 and IPv6).

3.3.2   Emergency Integrated Architecture

As anticipated, the future integrated emergency systems will be characterized by the utilization of meshed heterogeneous architectures, based on both satellite and terrestrial segments; the satellite and terrestrial components are defined with particular attention on the transmission of the localization and sensing information, even though for some services the satellite component may be not necessary, neither for communications nor for sensing.

In this architecture (Figure 3.2) the satellite infrastructure plays a lead role for its independence from the catastrophic event as well as for its ability of

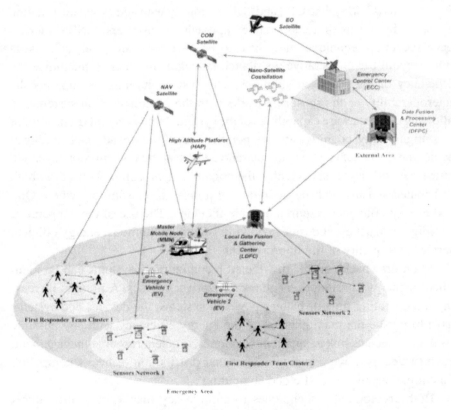

Fig. 3.2   Emergency system baseline architecture.

collecting information created by sensors deployed on the territory. This system covers the issues of sensing, security and safety information, through the collection of information via satellite and in-situ and the data harmonization capability.

The considered architecture enables an assisted localization, through the integration of GPS information and the information received from other in-situ terminals. The processing issues, required by the adoption of data fusion techniques for sensors and cooperative terminals in emergency scenarios, also play an important role particularly for data aggregation of sensors and terminals.

The sensors, terminals and interfaces used in this architecture aim to the acquisition and integration of biometric and geographic data from different places on the territory: they are highly re-configurable due to the large number of systems that could have be survived to the catastrophic event, include advanced localization functionalities and manage adaptive coding of source (video) and transport on multiple channel. The introduction of object domain deals with both sensors and teams as elements of the acquisition network. Assuming that pre-existent (to disaster) communication networks are totally or partially unavailable, by spectrum sensing techniques, spectral "holes" will be highlighted and opportunistically used for emergency communications.

3.3.3 System Architecture for e-Health Scenario

The definition of both the network architecture and the services platform solution for e-Health services derives from an accurate analysis of the user's needs. In order to perform the implementation of the overall e-Health infrastructure the following steps have to accomplished:

— definition and development of the services platform
— identification of the network requirements to support the defined services
— design of the network architecture satisfying the network constraints, exploiting different technologies characteristics and capabilities
— security and end-to-end QoS solution implementation.

3.3.3.1 Network architecture

The provision of e-Health services through the implementation of an interactive service platform, including real-time audio and video interactions among patients, specialists and health service providers, requires the deployment of an interconnected system based on an integrated and interoperable telecommunication network, which consists of both terrestrial and satellite segments.

Figure 3.3 describes the overall network architecture to provide the quality of service required by the considered e-Health applications, optimizing and minimizing the offered service cost. As shown in Figure 3.3, both citizens and physicians can access the interactive Service Platform from different locations (e.g., Health Points, Hospitals, Home) regardless of the chosen access technology, either satellite or terrestrial.

3.3.3.2 Service platform

The Service Platform (an example in Figure 3.4) aims at sharing health information among different applications and services (Self-Care and Assisted)

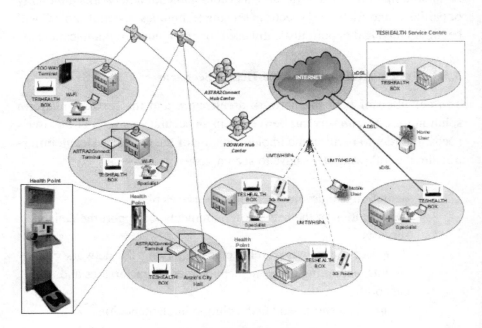

Fig. 3.3 e-Health network architecture.

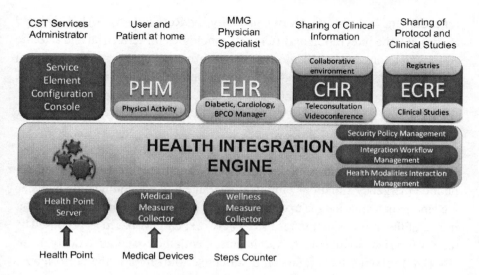

Fig. 3.4 Service platform for e-Health applications.

and it is based on the Health Integration Engine (HIE). This module guarantees the information exchange among the e-Health subsystems (Personal Health Media (PHM), Electronic Health Record (EHR), Electronic Clinical Research Form (ECRF), Clinical Health record (CHR)), creating work-flows and rules to acquire, manage and route health data. HIE emulator implements SOA (Service Oriented Architecture) paradigm and deals with all messages to/from the HIE compliant with the HL7 (Health Level Seven).

3.4 Services and Systems for Quality of Life

Some results are presented that have been already obtained in the field of Quality of Life improvement with specific reference to Emergency and e-Health applications.

3.4.1 Satellite-Assisted Localization and Communication for Emergency Services

For the emergency systems it is worth mentioning the main results recently obtained in the framework of the Italian National Research Project SALICE (Satellite-Assisted Localization and Communication for Emergency services), funded by the Italian Ministry of University and Research [8]. SALICE aimed at studying the integration of different technologies for terrestrial and satellite communications and localization in a single infrastructure by means of a digital platform implementation based on Software Defined Radio (SDR) technology. Some of the main achievements of the SALICE project are reviewed [9].

3.4.1.1 Cooperative relaying

Hybrid satellite/terrestrial cooperative relaying strategies are proposed for public emergency situations aiming at guaranteeing communication between the emergency area and the external areas [10]. As reported in [11], the combination of the hybrid satellite/terrestrial network, OFDM-based as proposed by the DVB-SH standard (SH-A Architecture), with the cooperative delay diversity (DD) relaying technique can be effective to overcome the performance loss in the non-line-of-sight environment. Moreover, cooperative relaying can be very effective in improving reliability and overall system performance: this could be especially important when the connectivity in a disaster area has to be restored and guaranteed.

In this cooperative scheme, the spatial diversity is mapped to frequency diversity to decrease the error rate: this feature achieves a remarkable performance improvement by the use of FEC codes. Both the satellite component (SC) and the complementary ground component (CGC) are included.

Results of computer simulations are presented in Figure 3.5, where the following working conditions have been selected: mode 1 K, signal bandwidth 5 MHz, OFDM sampling frequency 40/7 MHz, OFDM symbol duration 179.2 s, OFDM guard interval 44.8 s, QPSK modulation, turbo code (coderate 1/4). The channel propagation models are: Lutz model and TU6 model for the satellite and the terrestrial channel, respectively [12]. The one-relay and two-relay schemes are considered in an urban environment, which represents the worst case. The bit error rate (BER) performance is reported for different values of the delay in order to represent the impact of this value on the DD scheme; in addition in order to implement a more realistic case, a different power allocation between the SC and the CGC is assumed (unequal power): specifically, the copies of the signals coming from the CGCs (all types) are characterized by a higher power level with respect to the signal from the satellite.

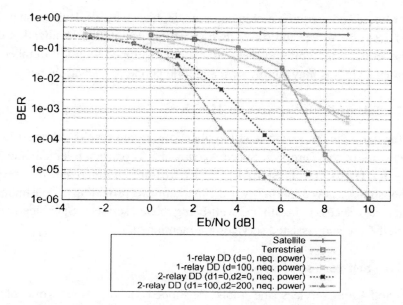

Fig. 3.5 DVB-SH performance comparison among: Satellite-only, terrestrial-only, one-relay cooperative DD and two-relay cooperative DD in city environment.

The presence of more than one relay, increasing the frequency selectivity of the channel transfer function, permits to achieve a significant BER performance improvement.

3.4.1.2 Cooperative localization

The use of hybrid positioning systems can increase accuracy and reliability by combining the pseudorange measurements of both GNSS and terrestrial ranging systems [13]. In emergency scenarios, the use of these cooperative approaches is mandatory, as the environmental conditions where the users' terminals are likely to be used can be extremely hostile for a standard GPS-based navigation receiver (indoor environments, urban canyons, operations under foliage, etc.). Assisted-GNSS (A-GNNS) systems improve the stand alone GNSS localization performance by using alternative networks to satellites (e.g., cellular network or the Internet) to provide the assistance data when and where no clear view of sky (e.g., light-indoor zones) is available. An innovative A-GNSS-like system, which does not rely on the existence of a pre structured communication infrastructure, is presented in [14]. It relies on a "peer-based" cooperation architecture in which each user's device (the aiding user) acts as a server and sends its own estimated satellite localization data to the aided user. While rescuers are usually equipped with GPS-embedded terminals, victims may simply own a mobile phone or a smartphone.

3.4.2   Telemedicine Services for Health

As for the e-Health services and applications, some results achieved in the framework of the European Space Agency (ESA) project Telemedicine Services for HEALTH (TESHEALTH) [15, 16] are summarized: this project demonstrates a new opportunity in the marketplace, providing people unprecedented easy access to health and well-being related services. Specifically both the Self-Care and Assisted services are mentioned.

3.4.2.1   Self-care services

For Self-Care services the tools developed in the framework of the TESHEALTH are the PHM and the Health Point.

The PHM is designed with the aim to help people to better know their body and adopt right life style. Acting as a tutor the PHM supports people in planning daily actions, thanks to an updated logbook dedicated to specific activities (e.g., check some parameters, make physical exercises, etc). In order to achieve the previous objectives the PHM includes are pository of care plans for primary prevention, which involves both the adoption of an alerting system (via SMS or email) for planned activities and a reporting system to make people aware of their results. In addition in order to allow people to share clinical information with the doctor, relatives and friends, it provides a documents and data storage. In detail, the PHM is a platform based on a three tiers architecture (Data layer, Back-End layer and Front-End layer). The Front-End layer represents the interface between user and the PHM system managing the graphical and textual messages. The Back-End layer hosts applications offering services to the Front-End layer. The Data layer is a database server used to store all data managed by the PHM application.

The Health Point is a remote device (a kiosk), that can be located in public buildings like pharmacies, allowing users to check autonomously some biometrics measurements. It permits to accurately measure: blood pressure, cardiac frequency, weight, body mass index and body fat. Consequently the cardiovascular risk evaluation can be estimated. These information are sent to the PHM in order to update the user health record.

3.4.2.2 Assisted services

For e-clinical studies the application scenario of these services is based on ECRF platforms and encompasses a multi-centric study, which includes a core lab as study monitor and several peripheral clinical centres as contributors to the study. Specifically, it involves the exchange of data among ECRFs (through the HIE) of a patient monitored in a multi-centric study between two hospitals. As an example, this permits to perform a study of a group of patients with cardiac chronical problems, by gathering a broad set of homogeneous data in a specifically dedicated database, which includes data collected during standard clinical exams. Patients enrolled in the protocol are followed up with periodical exams in hospitals and periodical controls, performed by operators at home. Collected data are transferred to the ECRF to perform controls and verify the protocol.

The ECRF application has three logical tiers:

— presentation tier: it implements the User Interface (UI), communicating with the Application and Data tier through secured Web Services and UDP for interactive multimedia. It includes the ECRF Web framework and the integrated collaborative video client
— application and data tier: it represents the back end of the application, including the Web and the collaborative video servers and also implements the application business logic
— HIE tier: it provides the bi-directional exchange of data between the ECRF application and the other TESHEALTH subsystems.

Tele-consultation services enable healthcare professionals to share patients data for early diagnosis and clinical cases discussion, allowing people to access periodical clinical checkup for prevention and medical institutions congestion reduction. A teleconference module includes a set of functions as datastorage and forwarding, collaborative working, virtual meetings, session scheduling, on-line video transmission and video-education.

Direct communication between users at different locations is provided. For example an expert could have a patient whose data need to be discussed with other centers. A virtual board enables all participants to exchange messages in a chat area and share images performing zoom actions and basic drawing as pointers, references, texts, highlighters.

3.5 Conclusions

The integration of communication, localization and sensing functionalities by means of heterogeneous and reconfigurable networks are a break through step for the growth of distributed cloud computing and social interaction technologies and a giant leap towards the provision of a plurality of services and applications, ideally and in perspective a universal library of services seamlessly provided to users.

Such new services can improve the quality of life of the citizens in everyday life. The two described services related to the tele-health and emergency fields are exemplary applications, where the flexible, multi-service and cooperative heterogeneous architectures play a fundamental role.

The integration of communications/localization/sensing functionalities in a heterogeneous, flexible, cooperative system is far from being presently available: the design, implementation and deployment of this visionary scenario to provide better services to the third-millennium citizens is indeed one of the most challenging issue for the ICT scientific community.

References

[1] European Commission Horizon 2020 — The Framework Programme for Research and Innovation, COM (2011) 808 final., Brussels, 2011.
[2] Emergency Communications (EMTEL) Requirements for Communication Between Authorities/Organizations During Emergencies, Std. ETSI Technical Specification TS 102 181, V.1.1.1, 2005.
[3] Chen C.-M., Macwan A., and Rupe J. "Guest Editorial: Network disaster recovery," *IEEE Communications Magazine*, vol. 49, no. 1, pp. 26–27, Jan. 2011. 10.1109/MCOM.2011. 5681010.
[4] Oberg J. C., Whitt A. G., and Mills R. M. "Disasters will happen Are you ready?" *IEEE Communications Magazine*, vol. 49, no. 1, pp. 36–42, Jan. 2011, 2006. 10.1109/MCOM.2011. 5681012.
[5] Del Re E. and Pierucci L. "An interactive multimedia satellite telemedicine service," *IEEE Multimedia*, vol. 7, no. 2, pp. 76–83, 2000. 10.1109/93.848435.
[6] Cova G., Huagang X., Qiang G., Guerrero E., Ricardo R., and Estevez J. "A perspective of state-of-the-art wireless technologies for e-health applications," *IEEE International*

Symposium on IT in Medicine Education, 2009, ITIME '09, vol. 1, 2009, pp. 76 –81. 10.1109/ITIME.2009.5236457.

[7] Zvikhachevskaya A., Markarian G., and Mihaylova L. "Quality of service consideration for the wireless telemedicine and e-health services," *IEEE Wireless Communications and Networking Conference*, 2009, WCNC 2009, pp. 1–6. 10.1109/WCNC.2009.4917925.

[8] Del Re E., Morosi S., Jayousi S., and Sacchi C. "SALICE — Satellite-Assisted Localization and Communication systems for Emergency services," *Proc. of 2009 IEEE Wireless Vitae Conference*, Aalborg (DK), May 2009, pp. 544–548. 10.1109/WIRELESSVITAE.2009.5172504.

[9] Berioli M., Molinaro A., Morosi S., and Scalise S. "Aerospace communications for emergency applications," *Proceedings of IEEE*, Vol. 99, no. 11, Nov. 2011, pp. 1922–1938. 10.1109/JPROC.2011.2161737.

[10] Morosi S., Jayousi S., and Del Re E. "Cooperative Delay Diversity in Hybrid Satellite/Terrestrial DVB-SH System," *Proceeding of the 2010 IEEE International Conference on Communications (ICC2010)*, Cape Town, South-Africa.10.1109/ICC.2010.5502626.

[11] Ben Slimane S., Li X., Zhou B., Syed N., and Dheim M. A. "Delay optimization in cooperative relaying with cyclic delay Diversity," *Proceeding of the 2008 IEEE International Conference on Communications (ICC2008)*, May 2008, pp. 3553–3557. 10.1109/ICC.2008.668.

[12] Digital Video Broadcasting (DVB) DVB-SH Implementation Guidelines, Std. ETSI Technical Specification TS 102 584 V1.1.1, 2008.

[13] Heinrichs G., Mulassano P., and Dovis F. "A hybrid positioning algorithm for cellular radio networks by using a common rake receiver," *Proc. Symp. Pers. Indoor Mobile Radio*, 2004, pp. 2347–2351. 10.1109/PIMRC.2004.1368739.

[14] Panizza M., Sacchi C., Varela-Miguez J., Morosi S., Vettori L., Digenti S., and Falletti E. "Feasibility study of a SDR-based reconfigurable terminal for emergency applications," *Proc. of IEEE Aerospace Conference*, 2011, Big Sky (USA), 5–12 March, 2011., 10.1109/AERO.2011.5747346.

[15] Kocian A., De Sanctis A., Rossi T., Ruggieri M., Del Re E., Jayousi S., Ronga L. S., and Suffritti R. "Hybrid satellite/terrestrial telemedicine services: Network requirements and architecture," *Proc. of IEEE Aerospace Conference*, 2011, Big Sky (USA), 5–12 March, 2011. 10.1109/AERO.2011.5747335.

[16] Ronga L. S., Jayousi S., Del Re E., Colitta L., Iannone G., Scorpiniti A., Aragno C., and Peraldo Neja C. "TESHEALTH: An Integrated Satellite/Terrestrial System for E-Health Services," *IEEE International Conference on Communications 2012*. ICC2012, Ottawa (Canada), 10–15 June, 2012.

Biography

Professor Enrico DEL RE was born in Florence, Italy. He received the Dr. Ing. degree in electronics engineering from the University of Pisa, Pisa, Italy, in 1971.

Until 1975 he was engaged in public administration and private firms, involved in the analysis and design of the telecommunication and air traffic control equipment and space systems. Since 1975 he has been with the Department of Electronics Engineering of the University of Florence, Italy, first as a Research Assistant, then as an Associate Professor, and since 1986 as Professor. During the academic year 1987–1988 he was on leave from the University of Florence for a nine-month period of research at the European Space Research and Technology Centre of the European Space Agency, The Netherlands. His main research interest are digital signal processing, mobile and satellite communications, on which he has published more than 300 papers, in international journals and conferences. He is the Co-editor of the book *Satellite Integrated Communications Networks* (North-Holland, 1988), one of the authors of the book *Data Compression and Error Control Techniques with Applications* (Academic, 1985) and the editor of the books *Mobile and Personal Communications* (Elsevier, 1995), *Software Radio Technologies and Services* (Springer, 2001), *Satellite Personal Communications for Future-Generation Systems* (Springer, 2002), *Mobile and Personal Satellite Communications 5-EMPS2002* (IIC, Italy, 2002) and *Satellite Communications and Navigation Systems* (Springer, 2008). He has been the Chairman of the European Project COST 227 "Integrated Space/Terrestrial Mobile Networks" (1992–95) and the EU COST Action 252 "Evolution of satellite personal communications from second to future generation systems" (1996–2000).

He has been the Chairman of the *Software Radio Technologies and Services Workshop (2000)*, the *Fifth European Workshop on Mobile/Personal Satcoms (2002)* and the *Satellite Navigation and Communications Systems (2006)*.

He received the 1988/89 premium from the IEE (UK) for the paper "Multicarrier demodulator for digital satellite communication systems".

He is the head of the Signal Processing and Communications Laboratory of the Department of Electronics and Telecommunications of the University of Florence.

Presently he is President of the Italian Interuniversity Consortium for Telecommunications (CNIT), having served before as Director.

Presently he is the Director of the Department of Information Engineering (DINFO) of the University of Florence, Italy.

Professor Del Re is a Senior Life Member of the IEEE and a member of the European Association for Signal Processing (EURASIP).

Simone Morosi (Member, IEEE) was born in Firenze, Italy, in 1968. He received the Dr. Eng. degree in Electronics Engineering and the PhD degree in Information and Telecommunication Engineering from the University of Florence, Florence, Italy, in 1996 and 2000, respectively. Since 1999, he has been a researcher at the Italian Interuniversity Consortium for Telecommunications (CNIT). Since 2000, he has been with the Department of Electronics and Telecommunications, University of Florence. His present research interests involve communication systems for emergency applications and future wireless communication systems. He participated in the European Network of Excellences NEWCOM (Network of Excellence in Wireless COMmunications), SatNEx (Satellite communications Network of Excellence) and CRUISE (CReating Ubiquitos Intelligent Sensing Environments).

4

CONASENSE as Cross-Cutting Challenge — A Dutch Perspective Based on IIP Intelligent Communication

Erik R Fledderus[1,2], Henk Eertink[3] and Patrick WJ Essers[4]

[1]*Netherlands Organization for Applied Scientific Research (TNO)*
[2]*Eindhoven University of Technology*
[3]*Novay, The Netherlands*
[4]*Seedlinqs, The Netherlands*

4.1 Introduction

Personal communication facilitates the exchange of information and emotions between (groups of) individuals. In most relevant applications, personal communication is facilitated and mediated by technology, e.g., computers, portable devices, sensors, internet, and (mobile) network infrastructures. Communication technology is, and will remain to be, a critical factor in many economic and societal sectors: for instance for health and well-being, entertainment and creative industries, mobility and logistics, and digital cities.

With the term "intelligent" we mean that communication systems will incorporate *awareness* of their heterogeneity and distributed nature, of their expected dependent and secure operation, and of their close collaboration with human users. No fully functional intelligent communication infrastructure exists to date. However, an operational intelligent communication infrastructure is tacitly assumed to be "ready for use" for most applications.

COmmunications- NAvigation-SENsing-SErvices (CONASENSE), 67–98.

The Dutch ICT community has brought forward a number of so-called ICT Innovation Platforms (IIPs). The platforms are meant to bring together researchers, entrepreneurs and 'users'[1] to develop a joint strategic research agenda (SRA). One such platform is the IIP Intelligent Communication (IIP IC). It seeks to bridge the gap that exists between demanded and the present intelligence in communication infrastructures. In this chapter, we highlight the commonalities between IIP ICs SRA and CONASENSE. IIPIC has explored three important societal domains to search for technological challenges and opportunities: health and well-being, smart mobility systems and smart energy systems. We shortly summarize the results below.

4.1.1   Health and Well-being

The Health and Well-being domain has been explored many times, but on the basis of this SRA research we identified three opportunities, that are aligned with other national and international agenda's (e.g., EIT ICT theme of Health and Well-being):

- The design and deployment of a health application platform for health related services, including an AppStore (and platform) for health services. Such services may support for instance telecare, with proper security and authorized access to relevant data.
- Contributions to safe and reliable communication across local (home and regional) communication systems, in order to ensure reliable connectivity but also with the possibility of prioritized communication across administrative domains in case of emergency situations.
- Improvement of the up scaling and uptake of new technology solutions within the health domain.

4.1.2   Mobility

The Mobility domain has recently received much interest, both in various European member states and cross-national initiatives. Interestingly enough

---

[1] We actually prefer the term 'requisitionary party', as a stakeholder that sets (non-) functional requirements and demands to solutions and their underlying technology. However, this term is quite cumbersome in a text, so we keep writing 'user'.

there seems to be hardly any concerted effort from telecom parties to address the challenges that originate from the proposed ITS solutions in this domain. The mobility domain is related to the (Dutch) top sector Logistics, and the EIT ICT Labs themes of ITS, Smart Spaces and Digital Cities [1]. Analysis of the mobility domain based on interviews with domain experts gives the following topics:

- Use of smart cognitive devices: their cognitive aspects allow them to easily join and leave a network. For instance within a car, between cars, between cars and roads, etcetera. Devices vary in capabilities (and power requirements). Low-power versions often use energy harvesting techniques. Others can be tablet-like or smart phones and require batteries.
- Recognition of mobility patterns using smart sensors (e.g., on smart phones). This can be used to provide insight in personal mobility patterns. This, in turn, will enable personal driving directions and even incentives for changing usual patterns in case of unforeseen road incidents, certainly when they are shared in 'private' social networks.
- Fusion of open data: what is the business model so stakeholders buy-in? What type of standards is required? We should also consider 'nice-to-have' services for end-users next to, for instance, professional traffic control services.
- Improving the scalability, safety and privacy of proposed Intelligent Traffic Systems solutions.

4.1.3 Smart Energy Systems

The added value of new communication technologies in the energy sector is very high. Electricity can only be stored at high costs — to the extent that today specific users are paid in hours of excess supply to consume electricity — and thus managing demand and supply is highly economical. This situation exacerbates when energy-supply increasingly decentralizes and the use of electrical cars would increase. The smart grid is the proposed solution to these problems. This is also one of the cornerstones for the top sector Energy and the EIT ICT Labs theme of Smart Energy Systems [1]. Part of the communication requirements for a future smart grid can be met with existing and future generally available communication infrastructures. However, several questions require attention:

- What will be the architecture of the communication infrastructure and management platform that supports the smart grids vision and how does it include various currently available infrastructures? How can the different reliability levels of different sets of data be managed, and how does the system deal with disturbances? The communication infrastructure and platform should also anticipate the application development by third parties.
- How should data from different sources be integrated? Questions refer to reliability of data transport and standards for data communication between energy providers, equipment providers and users, and between users. This also includes privacy issues and potential business model challenges when data needs to be shared among consumers, prosumers and grid-operators.
- How can the transition be organized between the current distribution networks and a future bi-directional smart grid from both a technological and business model perspective?

4.1.4   Cross-cutting Challenges

From the challenges above we can derive generic challenges for the 'intelligent communication' community. Behind each topic we emphasize the relevant part of the CONASENSE 'body of thought'. Key questions are:

- How to deal with reliability and prioritized *communication* over heterogeneous networks? (COnasense)
- What are requirements and technologies for open application platforms in order to effectively support third party applications and services (e.g., basic platform services and software-development-kits)?
- How to access, exchange, provide and control *data coming from various sources*, complying with standards while ensuring privacy and scalability? (conaSENSE)
- Significantly increase the availability of solutions in domains with strong availability requirements. Implementation and up scaling

of new *(communication) networking* and service technologies in domains may require new concepts. (COnasense)

After a brief state of the art of intelligent communication the remainder of the chapter elaborates on the three individual domains — intelligent communication, smart energy systems and smart mobility systems — and finalizes with an extensive treatment of the cross-cutting challenges.

4.2 Intelligent Communication

4.2.1 Introduction

Communication is the art of exchanging information. It is an essential ingredient of life, inside and between organisms, and it also takes place between devices that contain 'artificial intelligence' such as computers and intelligent sensors. It is achieved by a combination of 'access to' and 'transmission of' information across communication channels and networks. This started with the plain old telephone system (POTS), that) that provided global audio connectivity throughout the 20th century. This network lacked portability and mobility. These drawbacks were addressed and resolved by subsequent generation's communication technologies. The main exponent of the second generation (2G) wireless systems was GSM — it supported portable and mobile voice transport services, while UMTS (3G) added proper data communication to it.

While speed of wireless links increases with a constant rate (Edholm's law, see Figure 4.1), at the same time the volume of intelligent devices decreases with a staggering rate as well. The intelligent devices of the 70ties and 80ties, mainframes, minicomputers and PCs, were connected with the emergence of Internet. Remarkably, the way computers communicate using Internet solves the second problem of digital communication: ubiquitous access. A computer is permanently on-line and has access to the whole world via the internet. 3G, and its successor 4G, combine the best of both worlds and allow connectivity at any place, at any time and for any type of medium. 4G intelligent communication solutions support end-users, groups, organizations in the specific context of their daily life and operations by immersing them in an information society. Connectivity, or the possibility to communicate, has become the key enabling factor for all forms of economical, societal, political, or educational relationships and activities.

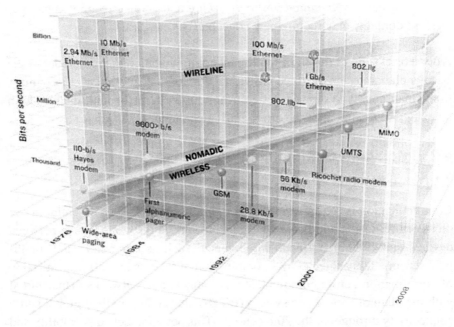

Fig. 4.1 Edholm's law illustrates how the telecommunication technologies march almost in lock step: their data rates increase on similar exponential curves, the slower rates trailing the faster ones by a predictable time lag [2].

New communication technologies are rapidly changing society. Individuals are increasingly immersed in a world with all sorts of devices that can interact with them; e.g., smart phones, displays and home stations. These devices should understand the user's information needs and act accordingly. The result is an ambient intelligent communication environment brought about by a combination of sensors, applications, devices, software services and communication technologies (see Figure 4.2 for the projected traffic growth that will result in). People and applications will communicate with each other through the appropriate networks — fixed or mobile — which are seamlessly intertwined.

The impact of intelligent communication is that users will have a stronger sense of control over their own life, and this will be irrespective of whether they are at work, undergoing medical treatment, traveling or interacting with agencies, or in a social context with family or friends. A patient, for example, will become less dependent and will enjoy a better quality of life in his own, trusted

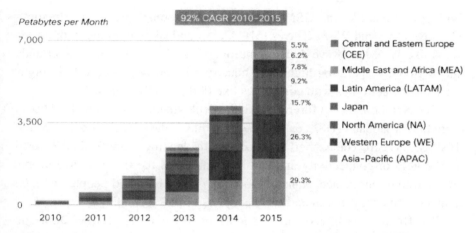

Fig. 4.2 A graph showing how traffic is expected to grow throughout the coming years, and distributed geographically. 'CAGR' is 'Compound Annual Growth Rate' [3].

environment. A police-officer will feel more secure in unfamiliar situations, being in contact with and monitored by his colleagues in the vicinity and back-office. Focused social interaction between (groups of) people will reduce loneliness and improve inclusion, social cohesion and well-being.

To realize this goal of being 'connected anytime, anywhere, anyhow' many issues need to be resolved. For a continuous Internet connection we probably need to rely on multiple access networks, such as WLAN, UMTS, GPRS, Bluetooth, etc., irrespective of whether we are a priori known to these networks or not. If multiple networks are available, the device that mediates our communication should use the one that is closest to our preference, e.g., the one with the highest bandwidth or with the lowest costs. But if we change network environment during travel, we do not want to change network settings or restart applications. The network should therefore know who we are, what services we want to use, what our preferences are, and who should eventually receive the bill. But this information will be continuously changing. In these dynamic settings, network and service providers should make a joint effort to guarantee a sufficient level of quality, security and privacy.

4.2.2 Communication Technology and the Netherlands

In the past, the Netherlands has played a significant role in the development of technologies for wired and wireless communication networks. For wireless, with inventions like the GSM speech coder (Kroon-Sluiter [3]), Bluetooth (Haartsen [5]) and Wi-Fi (Hayes [6]). And in optical networking with inventions like WDM — wave division multiplexing (Smit [7]). The Netherlands has a globally operating high-tech industry with an important role in high-frequency electronics with companies like Philips, NXP and ASML.

The Netherlands is a forerunner in the deployment of high-end backbone, wire line broadband (DSL, cable, fiber) and in wireless 3G coverage (UMTS, HSPA). As such, The Netherlands is one of the leading countries in the world in the field of connectivity and can be regarded as the Gateway to Europe. Furthermore, public acceptance of internet is very high and people are open minded, efficiency driven and early adopters of new technology.

The Dutch services economy accounts for 70% of the total Gross Domestic Product (GDP). Almost the entire growth in employment during the last decade stems from developments in the service economy. Sectors such as media, finance, business services, health and service/product combinations are pillars of our service economy. Being a service economy means that the development focus for communication infrastructures will be influenced by the support of the communication infrastructure for specific application domains, e.g., in healthcare, media, public services, finance and education.

In 2001, a consortium of industry and academia started an initiative in the Netherlands to create a shared R&D vision that underlines the need for the transition to a more flexible communication infrastructure supporting the needs of the Dutch economy and society. This resulted in the 'Freeband' vision. This vision on telecommunications described how in 2010 consumers and companies will have unrestricted access to information and communication, through unlimited bandwidth and highly sophisticated, personalized ICT services. As a result, the Freeband program started with the goal of raising the level of knowledge on telecommunications at academia in the Netherlands to an international top-level.

4.2.3   Current Status in the World

4G is a logical evolution from 3G (UMTS) and fixed broadband networks towards an integrated system. While 4G is currently not realized as a whole, many technological and business developments suggest that 4G, and in particular Long Term Evolution (LTE) is well underway to become the new reality:

- Broadband xDSL or Cable connections up to 100 Mb/s transmission rates are deployed at a staggering rate in developed countries. In the Netherlands, we have a high penetration of broad band connection in the Dutch end-user market. Substantial FttH deployments will significantly increasing the average data rates over the coming years.
- Wireless Wi-Fi routers deployed in households are common. These offer data rates often above 54 Mbit/s, with 802.11i security providing perspectives for increased protection. Wi-Fi data rates are boosted to 500 Mb/s (802.11 n) and even 1 Gb/s (WiGiG [8]).
- A new generation of integrated devices with extended viewing facilities and ubiquitous internet access are hitting the market. Such devices allow people to be connected anywhere, any time and with the best possible means. Typical examples are the smart phones and tablets that open-up the post-PC world and will change the way we experience successful desktop-pc services, such as internet telephony, RSS, infotainment, or social networks. The winners in these new markets will be those who can integrate functionality in the most appealing and intelligent way.
- Unlicensed Mobile Access (now known as Generic Access Network, GAN) support seamless roaming and handover between local area (Wi-Fi) networks and wide area networks such as UMTS using the same mobile phone. First devices have appeared on the market that supports these features.
- Telecom operators offer subscriptions that allow users to use their mobile phone in home as a normal phone, based on so-called Picocells (Wi-Fi routers that contain an additional 3G radio, connected via the home access connection to the selected 3G or 4G mobile network). This is completely in line with 4G fixed-mobile convergence: an all-IP, user-centric philosophy.
- Triple play telecom services that combine TV (media), voice, and internet browsing are becoming common. Quadruple play (including mobile services) is starting off.
- Blogs and content-sharing websites: global sharing of personal content like photos and videos has become possible with sharing websites of YouTube, Facebook and Flickr.

- The Internet is the communication and networking medium of choice. However, the Internet itself was not designed for this. New bandwidth efficient and liberal technologies for sharing information have emerged. P2P communication has invaded the internet as the main use case. This accounts for over 70% of traffic nowadays (June 2011), largely surpassing emails, web browsing and other applications. While at first mainly used for illegal sharing of content, these technologies are now being applied for mass-scale internet-based communication services. These technologies offer the promise of efficient use of network resources, essential e.g., for next generation television broadcasting. Examples are Skype, Joost, BBC-iPlayer and Tribler [9].

4.3 Health and Well-Being

Healthcare is one of the most pronounced societal areas facing large challenges, involving patients, citizens, government, healthcare institutions and companies. ICT is considered an important enabling technology. Health and wellbeing is a broad domain. It is often broken down into:

- *Cure.* This is the domain of hospitals: diagnosis, surgery and care are the main processes. From the IIPIC point of view there are challenges in the management of patient records, person-to-person communication, and wireless communication in surgery rooms. Treatment in hospitals is, in The Netherlands, funded by health-insurance companies on the basis of a well-defined treatment.
- *Care centers.* This is the domain of elderly and people that require a high level of care. These are financed by the government by means of the AWBZ (a Dutch abbreviation for an *Exceptional Medical Expenses Act*). The core business of care centers is to organize (personalized) 24x7 care services for their clients. There is a move towards decentralized care: smaller care-centers in neighborhoods, often requiring remote observation of inhabitants during out-of-office hours.
- *Telecare service providers.* These organizations provide services like vital-sign monitoring of individuals, connected either synchronously or asynchronously with a telecare center or hospital.

Other variants involve smart homes; the in-home technology is connected to a telecare center and/or a general purpose alarm-centre.
The business models for telecare are not really established yet. Telecare service providers are not always associated with a care-center;
they can also be an intermediary between multiple care providers.
Services are often tied together with specific hardware solutions,
resulting in minimal interoperability between in-home solutions.

- *The care-chain.* Professional health and care centers have to cooperate: patient-centered care requires the combination of work-
flows across organizations, in combination with the exchange of
the relevant medical data records. Concerns are process optimization, information security and privacy, efficient exchange of high-
resolution images, and cost efficient operation.
- *Lifestyle, Prevention, and Well-being.* Prevention is both primary
and secondary prevention. Ultimately, this is the domain that is
driven (and paid for) by the consumer. Example services are support for fitness, cycling, or running, assistance for cooking, stop
smoking, lifestyle coaching, etcetera. Although some of these services are (co-)funded by employers, often they are paid for by consumers (e.g., via downloaded apps on smart phones in combination
with a service-subscription).

4.3.1 Visions on Health and Well-being

Health and Well-being visions are widespread. We quote one of them:

- European Foresight Network [10]: "(Social) informal care is a
paradigm to stimulate the wealth of vital, social relationships [11].
New social societies bloom in the vicinity of community centers,
in city quarters and districts. Internet technology, especially that
which constitutes a semantic web [12], enables virtual communities
to form and operate. Voluntary, charitable care must be nurtured,
because it is indispensable for encouraging autonomy, self-esteem
and eloquence in the elderly. The opportunities and advantages of
informal care by family, friends and social networks particularly
should receive more support."

4.3.2 Domain challenges

In Table 4.1 we give the prominent domain challenges, resulting from a number of workshops and interviews.

In addition, the special character of the health and wellbeing domain puts special constraints on innovations and technical solutions. Boundary conditions to solutions are:

- *Ease of use*. Care puts heavy demands on ease-of-use, because of

    o users of technology vary a lot in skills, and

    o technology is being deployed in critical circumstances, e.g., in operation rooms or in emergency situations, and should therefore be very intuitive

- *Robustness*. System failure is not acceptable because it might lead to loss of life or loss of quality of life.

Table 4.1.   Challenges for the health and well-being domain.

Domain challenge	Description
Waiting lists	Societal most pronounced and commonly known challenge: the need to wait before treatment can occur in hospitals, health centers or at medical professionals. This leads to the emergence of private hospitals for specific treatments.
Lack of personnel	Especially in care: nursing professions and informal care. We have an ageing society, which leads to a higher demand for care in combination with a reduction of the work-force.
Lack of resources, e.g., expensive MRI scanners	Cost of procurement, duplication of equipment, non-optimal scheduling, waiting lists
Dealing with increased availability of new (costly) treatments	Due to scientific developments, the number of possible treatments increases. Potentially a multitude of patient-centric and disease-specific applications, services and solutions is needed.
Fragmentation of care organization	• Market is difficult to penetrate for newcomers, also due to regulatory issues; • High entry costs–economies of scale difficult to achieve due to fragmentation; • No clear problem owner, inhibiting the uptake of innovations; • Hard to keep track of overall 'patient' process.
Financing structure	No clear problem owner, inhibiting the uptake of innovations.

4.3.3 Scenario 'Health and Well-being in 2025'

Around 2025, care will beorganized differently. Because of the continuously increasing number of treatment methods (with expensive technology), the costs of healthcare have surpassed all boundaries of the collective affordability. Health and treatment has in fact become a scarce commodity, so that economic laws put a much stronger marker on the organization of care than in the past.

Specialized centers, both public and private, play a major role in the daily care as well as in health counseling. This is both a direct consequence of efficiency targets and market forces. On the other hand it also has a societal cause: people with health & wellbeing wishes and demands (formerly denoted as "patients") want optimal care, and are no longer satisfied with an average 'treatment'; this trend is further supported by increased mobility of people. Therefore many people are turning to specialized health centers. These centers (large and small) are often the result of cooperation between traditional providers and private organizations: insurance companies, employers and care providers (e.g., fitness centers, gyms and housing associations) that other than strictly healthcare-related reasons, invest in health and wellness.

The strict line between sick and healthy has disappeared. It has been replaced by a continuum of life-long well-being, in which an individual and his next of kin are basically responsible for seeking and addressing the right care at the right time. The market offers numerous tele-medicine or self-care solutions that meet the wishes and necessities of people to monitor their own health, in the most pleasant way possible.

Use case scenario: Somewhere in the not so far away future…

Marijke loves to exercise. In her 47th year she still feels pretty fit, and spends considerable time at her work as a casting consultant. Still, life has left some traces. As a result of thrombosis with pulmonary embolism a few years ago, her lung capacity is reduced, and she is forced to swallow anticoagulants. But especially for her an active lifestyle is recommended if she wants to stay as healthy as possible. But her condition has made Marijke a tad uncertain. She keeps a close look on her heart rate and blood pressure during exercise, even though that does not directly relate to her lung problem.

Her gym reacts to her situation through its own portal by offering several health services that visualize various health aspects. The data from the Nike App on her iWatch are directly visible on the portal (which is why she wears the watch all day, precisely because of this functionality), but also the details of her last workouts including the response of her body to the training efforts. She has set the alarm somewhat sensitively, so that there is timely warning if her heartbeat shows deviations. Given her past health issues, she has also arranged for a warning signal to telecare-center in case of a heartbeat disorder. Because she knows that the web-portal has been developed in conjunction with a medically-recognized consulting firm, she knows that the portal is more than just a few inaccurate graphs. What's particularly nice is that once a month, the doctor at the sports center performs an extra scan of her data, to see if there are any peculiarities to be noted. For which there is no extra charge by the way.

He grabs her racket and a squash ball which is still cold. Then the phone rings. She sees immediately that the ringer is one the informal care-takers of her single mother, but from the green color of the screen she already knows that there is no alarming situation...

4.3.4   Research Questions for the Health and Well-being Domain

4.3.4.1   Requirements

- The home network should be extensible. This home network should be future proof in order to support new applications, including robotics, or health support at home.
- The platform should support personalized user interfaces. How to design personalized user interfaces for people/patients?
- A platform should interface to social network sites, Instant Messaging, and e-mail
- A unified integrated health log should be available to patients. This should support the addition of new types of information, coming from multiple sources, in various formats and through a variety of channels.
- Resilience against failures. If the network fails critical messages (alarms, medication reminders,) should still be given. Fall-back scenarios (e.g., via public mobile networks) must be available.

4.3.4.2 Technical challenges

- Standardized data formats and efficient lower layer transport protocols.
- Ensure that sufficient bandwidth is available.
- Define standard programming interfaces for services, that can dynamically switch between lower layers, and that automatically discover the most optimal network.

4.3.4.3 Data storage

- How to deal with scalability of 24/7 monitoring with respect to data storage — should all (raw or processed?) data be stored — if so, for how long? Discard raw data and keep only processed data or high level interpretations/abstractions derived from that data? Thus losing the audit trail…
- How to select appropriate data or decide which data to be discarded or not; whichever choices are made, legal and privacy implications, as well as potential medical consequences, can follow.
- How to identify and use patterns in large amounts of sensor data?

4.3.4.4 Solution directions

- Define a virtual care network as an overlay over existing infrastructure.
- Create a Care Exchange: an application layer switch between (smart) homes and care centers.
- Unobtrusive monitoring using various technologies: For use in the extramural case this is a must to get it broadly accepted. This means reliable (i.e., handling artifacts), small (i.e., integrated), easy to use (i.e., battery-less, hidden in textiles) and low cost (i.e., ready for the mass market).
- Can we use Service-oriented computing as a solution?

4.3.4.5 Business models and innovation

- How to orchestrate care around a particular patient? Can this be related to the care-exchange from the previous paragraph?

- What are usable go-to-market strategies for in-home care solution providers?
- How to design new healthcare processes that reflect the available and upcoming technological possibilities?

4.3.4.6 Authorization, authentication and privacy

- How to arrange multi stakeholder access to information?
- How to guarantee the privacy of heath related data?
- Storage and retrieval services for medical (sensor) data. What are appropriate search algorithms? Data filtering and diagnostic recommendation

4.4 Smart Energy

Sustainable energy production and consumption are critical components in dealing with climate change, CO_2 emissions and exhaustion of natural resources. The load on the energy network is increasingly unpredictable due to decentralized energy production (wind mills, solar cells). That's why in particular the power network needs to be upgraded to a smart grid: our energy system turns into a smart energy system.

Smart energy systems support controlled charging of electric cars, deploys smart grids with local, domestic energy production, energy delivery and usage. This converts the electricity network into a high tech system which relies heavily on information and networking technology.

4.4.1 Visions on Smart Energy

Smart Grids represent the cutting edge of energy efficient technologies, applied in energy production, distribution and householder use. Smart grids are modernized electricity grids that interact with information technology and communications infrastructure to provide greater transparency on energy use to consumers, and to improve the efficiency of energy supply. Smart grids support the integration of renewable and distributed energy sources into the grid, like solar, wind and co-generation plants. Smart grids promise to be more reliable, with fewer and shorter blackouts. They allow electric vehicles to be charged when demand on the network is low, and their combined battery storage can be used to support the network when demand is high. Consumers need no longer be passive receivers of power, but instead can take responsibility of their energy use and make meaningful decisions that will benefit both the environment and their hip pockets.

From the perspective of commercial energy producers the most important aspect is peak-shaving: shifting energy consumption to periods with less energy demand. This can be done by providing incentives to consumers, or by remotely controlling household appliances. Many of these changes will help lower peak electricity demand and reduce the need for electricity companies to build extra power plants simply to cope with daily peaks in consumer demand."

Key elements of this vision:

- Production of green ('renewable') energy. This can be done by commercial energy producers, but also by individual companies or households (solar cells, wind-mills).
- Support for electric cars & chargers. This leads to new local power demands on one side, and significant electricity storage on the other side.
- Smart Metering Infrastructure (SMI) and in-house controls, to provide insight in energy usage for inhabitants.
- Advanced tariff differentiation, to provide incentives to increase or reduce current power consumption.
- Information exchange in the energy supply chain, in order to predict energy usage and control energy production.
- Communities may support local trading of energy
- Local E&G/W&C projects: specific solutions not related to the main power grid (especially local Warmth & Cooling solutions)

4.4.2 Domain Challenges

These challenges are based on interviews and a careful analysis of existing roadmaps [13–16] — see Table 4.2.

Extrapolation These challenges are likely to become more prominent in relation to the increasing energy demands and requirements for sustainability.

Boundary conditions to solutions:

- Use of "proven technology". Extensive roll out of new technologies in experimental settings is needed before a national smart grid is realized.

Table 4.2. Challenges for the smart energy domain.

Domain challenge	Description
Supplying the market with sufficient energy without building energy plants	• Home automation to reduce energy-waste. • Create Smart Grids to more accurately predict periods with high or low demands. • Influencing user behavior: much of the demand is still user-induced
Creating an infrastructure for electric cars	With sufficient charging stations capable of supplying the required energy
Delivering clean or pure power	Maintaining the "quality" of energy production and distribution.
Storage of energy	It is often cheaper to produce energy than to store energy.
Delivering richer services to clients	Energy is a commodity, utilities may exploit the 'smart meter' to deliver added value services

• Long time horizon: energy grids have an economic Return of Investment (RoI) of tens of years

4.4.3 Scenario "Smart Energy in 2020"

The energy market is highly diversified. Many sources of electricity are used in combination (traditional sources like gas and coal, wind, water power, decentralized solar power, decentralized small energy plants). The supply mix and demand mix is fully automated, based on rules defined by users and energy suppliers. In other words, based on weather predictions, demand predictions and user rules decentralized systems decide whether it will load electric cars, adapt lighting on high roads, or adapt energy production in gas power plants. Energy companies are competing on price and intelligent customer services, like fully automated control of household equipment based on energy supply and prices and user preferences. The supporting ICT infrastructure in the home consists of fine-grained communication networks, which exchange (aggregated and/or anonymous) information with the energy company and various third party service providers via the Internet. Users become more aware of the variety of possibilities for utility services including home-specific solutions for electricity production and saving, heating and cooling, cooking and water management. Energy-consumption can be remotely managed and controlled through web-interfaces, on the desktop, touchpad or mobile. Support for energy equipment is managed by the energy company, using a supply base of local professional heating service companies.

Use case scenario: Somewhere in the not so far away future...

I get up early this morning: today we have to decorate the living room since it's my daughter's birthday! It's mid-winter and cold outside, but luckily the heating remembers this special day from last year. It's already warm in my bedroom and in the living room, while the other bedrooms will still be cold. Nice system. My only concern is all the personal data that my energy company seems to collect. They give nice privacy guarantees, but lately it was in the news that an energy company, not mine, had kept the data after the client had moved to a competitor, sending birthday cards to his wife reminding her of the nice time they had when still a client. And I still do not get why apparently it is beneficial that my microwave is web-enabled. Anyway, my invoice was lower again, particularly since cloth washing is done on last-minute low-cost hours. Just filling the machine in the evening and the wash is clean in the morning. Oh god, I still have to get it from the machine...

4.4.4 Research Questions

4.4.4.1 Smart grids

- Should the electricity grid manager(s) play a role in the home-to-grid(s)-to-producer smart grid communication?
- How can a smart grid deal with all parties involved in its operation?
- If users make different types of agreements with energy providers on the terms of the contract (time of delivery, service levels), how can the electricity grid manager guarantee sufficient capacity?
- How can we model and simulate Smart Grid Systems? Which modeling techniques exist for energy provision, in which the ICT system is incorporated? How could we use perfect knowledge on the system?
- Should the smart meter only register energy consumption and production, or also control appliances?
- Is energy supply sufficient for electrical transport? Can we manage scarcity?
- How can the interests of stakeholders be represented by intelligent agents, so that they can jointly negotiate their mutual interests?
- How do you communicate energy use to users in a comprehensive, yet attractive, fashion?

4.4.4.2   Communication technology

- Which communication technologies are needed for the smart grid? Does fiber technology offer real advantages? Is the use of the power grid itself an option? Do we need one infrastructure, or several ones?
- How can ICT facilitate the efficient use of energy transport capacity? Is it possible to determine optimal loading of (connected) parked electrical cars?
- Which information do we want to communicate? Which part of the information is time-critical?
- How can we transport sets of data with different reliability levels within the same communication system? Can fault-tolerant systems contribute?
- How can the life span of an energy system (long) and a communication infrastructure (short) be aligned?
- How can in-house information be transported with a sufficient degree of reliability, without new wiring? Is the electricity network appropriate for this purpose, even if this would also be used for other purposes? Could it be wireless?
- What is the transition trajectory from the present state to a communication system for a smart grid? Is there a standardization trajectory?

4.4.4.3   Conditions for implementation

- Business models: How can a supplier of communication services generate revenues and profits by transporting small quantities of data with high levels of reliability?
- Privacy: how can we guarantee user privacy in a smart grid?

4.5   Smart Mobility Systems

Mobility of persons and goods is a key asset in modern society; in order to preserve and enhance mobility new mobility solutions are needed that guarantee both safe and sustainable 'throughput', in an environment-friendly manner.

Wikipedia states: "Mobility is the ability and willingness to move or change". In practice, mobility is associated with various aspects

- Logistics — transport systems,
- Mobility of persons: being able to move around, taking a personal environment 'along'
- Tracking & tracing: tracking objects, RFID technology, virtualization of physical objects

4.5.1 Visions on "Smart Mobility Systems"

Typical visions are based on ITS and personal mobility:

- ITS (Intelligent Transport Systems): Navigation systems will develop into a full-fledged information system, with which the user has access to information about the traffic situation to be expected, can reserve and pay for parking places, can be given advice on his route and destination, all tailor-made for his situation and preferences. The car communicates with systems on the road and systems in the car. These systems can warn the user of special or dangerous situations and help the user to take action. In special situations, such as traffic jams or in emergency situations or on special lanes, the car will take over the driving from the driver [17].
- Personal smart mobility: this exploits the availability of multi-modal, real time, and enriched travel information. A traveler with a journey from 'a' to 'b', gets provided with modality suggestions (public vs. private transport) real-time information on deviations, congestion and other irregularities, relevant context information and extended service options such as reservation, billing and ticketing or presence information.

More specific trends visible in visionary statements are [18]:

- Intelligent cars/vehicles. This is enabled by the concept of 'Connected cars': car-to-car communication, car-to-roadside communication
- Automated Traffic guidance

- Multimodal route planning: integration of public transport in navigation systems
- Intelligent routing (in particular in the logistics sector)
- Congestion management: pay for use, incentives to change standard mobility patterns ...

4.5.2   Domain Challenges

In the Table 4.3 we give the prominent domain challenges

With the increase of people living in cities, all three challenges will only become more prominent. Together they form important cornerstones of a livable city. The suggested technical directions make clear that some important boundary conditions need be obeyed:

- Availability and reliability of information, in relation to potentially limited connectivity via mobile networks
- A sufficient level of inherent privacy of user generated data; a sufficient level of open access to data to stimulate service innovation and service eco-system
- Scalability of small-scale solutions

Table 4.3.   Challenges for the smart mobility domain.

Domain challenge	Description
Towards a sustainable mobility system, for people and goods	Moving people and goods takes its toll: in terms of fossil fuels, noise and particulate matters. Can we reduce unnecessary movements, or replace them with more sustainable ones? What are the commercial, legal and behavioral thresholds?
Reducing traffic jams, increasing system throughput	The average fill rate in logistics is 45%, the average car-to-car distance in the Netherlands lies below the average European distance (2 seconds). What is required to further increase the throughput? How do the various aspects (human in the loop, technology, commercial/legal) contribute to the current status quo?
Safety	Roads get more crowded, and are used by multiple modalities (pedestrians, bikes, cars, buses, trucks ...). Reduction of casualties is a key ambition of many European countries, and directions involve both technology as well as the human in the loop. For instance, car-to-car and car-to-roadside communication could assist in preventing accidents to occur.

4.5.3 Scenario "Smart Mobility in 2020"

Twenty years ago, the average Dutch citizen viewed traffic jams as an unsolvable problem, or even as a characteristic of the daily routine. This perspective changed radically in the second decade of this millennium. The possibility to choose reliable alternatives at any moment of a journey persuaded even the most convinced car owner. A recent crowd-sourced investigation by TNS NIPO showed that 80% of the respondents use on a daily basis at least two traffic modalities. Moreover, 50% indicated to use three or more! The most important drivers are the reliability of predictions of ToA (time of arrival) and the punctuality of the various public transport alternatives. The high percentages in the TNS NIPO investigation are remarkable because of the recently introduced cooperative driving mode on the main highways in the Randstad. Cooperative driving has improved the throughput (car-to-car distance decreased to 0.7 seconds) and the reliability, and has decreased the chances to end up in a traffic jam. In a twin-interview with the president of the ANWB and the minister for Mobility and Logistics, the former indicated that the tax on personal CO_2 footprint should be viewed as the most important reason for this change in behavior. This tax addresses sustainability in the widest sense of the word and leapfrogs the controversial and hotly debated road pricing proposal from 2009/2010. The minister proudly mentioned the European award in the area of 'sustainable regulation', a price for implemented policies and regulations that stimulates sustainable behavior.

Use case scenario: Somewhere in the not so far away future…

It's Saturday, and Melanie wakes up with a smile on her face. Ever since the daily stress for traffic jams vanished, and her last resort-telephone calls to Harm to pick up Claire from OSC (out-of-school care) were no longer needed, Melanie can start her weekend more refreshed. And today is even more special — the whole family will have a day out at the Deventer book market. Her 6-year old son Pim works already for some days on his tablet to prepare a suitable travel scheme. This month at school is dedicated to 'sustainable mobility', and the various groups compete with their personal CO_2 footprint for a mouth-watering price: a long weekend in 'the house of the future' in Almere, including a survival in the Oostvaardersplassen. Melanie indicated that she wanted to combine sustainability with relaxedness, and since Pim's sister Claire is only 4 years old, he should take care of a number of constraints …

Two hours later, Melanie is enjoying a lovely ristretto in the newly build Utrecht CS Experience Centre — the Centre has been finished recently as the first part within the CU2030 plans of the city, and has already been nominated as the largest multi-modal interchange station in Europe. Pim and his father have been able to combine their personal all-in travel-voucher with some nice surprises: a reduced price for the ristretto and for every member of the family a free e-book. Claire is playing with her 3D version of Cinderella. However, close to Apeldoorn the excellent mood disappears: the driver-avatar appears on the screens in the back of the seats, and indicates that the journey will come to an abrupt end. A cooperative driving vehicle on the loose has created havoc on one of the nearby crossings. Luckily, the Dutch railways are able to quickly provide an alternative: taxi-coaches. However, Pim objects to this alternative, since he will lose too many CO_2-goodies... Using his tablet he finds a better plan B: the high-speed bus on the A1 makes an extra stop at the Apeldoorn-interchange, and will continue directly to Deventer. With a delay of only 5 minutes the family meets with the rental-bike owner in Deventer, and two minutes later they are cycling on their fashionable bikes to the book market...

4.5.4   Research Questions

The most important research questions are categorized as follows:

4.5.4.1   Realizing scalability

- How to use spatial selectivity (e.g., using position dependent beaming with multi-antennas) to improve the efficient reuse of resources
- How to manage distributed systems of variable size?
- Striking the balance between ad hoc and infrastructure based solutions
- Efficient use of spectral resources for tracking and tracing
- How to integrate the mobility and logistics domain with the IP domain?

4.5.4.2   Sustainable solutions: low-power/long life time

- How can we develop ultra-low power solutions for positioning (RFID and ultra-low cost tags)

- How to develop intelligent tags, with more info, enabling two-way (send/receive) communication?

4.5.4.3 Realizing trust and privacy

- Trust to enable domain-crossing cooperation and service development
- Trust to authorize proper access
- (Inherent) privacy to stimulate on the fly aggregation of information

4.5.4.4 (Technologies enabling) system innovation

- Charting market imperfections and unwanted control points that stale innovation
- Realizing an inventory of standards and systems that could enable the various services
- Data fusion for large-scale sensor systems
- Towards cooperative driving
- How to take care that vehicles are able to drive closely together at high speed?
- How to use broadband mobile (RF, wireless) communication for inter-car communication, anti-collision communication, radar, …

4.6 Cross-Cutting Challenges

4.6.1 Communication …

Communication is the exchange of information between (groups of) individuals. In an increasing number of domains, communication is facilitated and mediated by technological means such as computers, portable devices, sensors, internet, and (mobile) network infrastructures. Communication technology is and will remain to be a critical factor in many economic and societal sectors, for instance health and well-being, entertainment and creative industries, mobility and logistics and environment and climate.

…is pervasive …

Rapid developments of mobile and internet technology in the past ten years have led to evident changes in the way society interacts and business is done. Communication and computing are everywhere, from television sets to navigation systems, and from web2.0 to sensors in fashionable sportswear. The average modern citizen uses tens of communication devices on a daily basis, often without realizing it. Thanks to embedded systems and wireless networking, we are reaching an era where technology is gradually disappearing into a pervasive fabric of communication utilities.

...yet is facing complexity challenges ...

Communication technology has long aimed at solutions for connectivity anytime, anywhere and anyhow. In sectors that are historically communication oriented, such as managed telecommunication and internet systems, the abundance of aptly called 3G-solutions (WLAN, UMTS, Bluetooth) gives an adequate answer to their connectivity needs. However, in newly emerging and innovate sectors — especially those that are emphasized as drivers for society and economy in the beginning of the 21st century — communication is still facing enormous complexity challenges hampering innovation across these sectors.

...for three reasons:

1. First, interaction between communication systems and its users is complex because it requires careful optimization between users, providers, applications and the autonomy of the supporting communication system.
2. Second, communication technology no longer connects only people, but also devices. For example the "Internet of Things", a wireless and self-configuring network between 100.000 billion objects such as appliances, personal devices, and clothing. Or sensor networks for collection health, energy, or environmental data. The sheer complexity of these heterogeneous communication systems makes it impossible to get away with today's ad-hoc solutions, but a structured architecture and design is required.
3. Finally, the behavior of these very large scale heterogeneous systems is hard to predict and therefore hard to control — especially in those cases where the communication architecture is decentralized, such as in peer-to-peer and ad-hoc networks, and in networks of embedded devices. Yet these systems should be scalable and operate dependently and securely.

4.6.2 Intelligent Communication ...

With the term "intelligent" we denote communication systems that incorporate awareness of their heterogeneity and distributed nature, of their expected dependent and secure operation, and of their intimate collaboration with human users. The adaptive, intuitive, and robust (AIR) technology that is needed to achieve such intelligent communication differs radically from what is provided by state-of-the-art solutions since these often rely on fully managed and centralized network architectures. No fully functional intelligent communication infrastructure exists to date.

...is needed ...

However, an operational intelligent communication infrastructure is from a technical point nearly always silently assumed to be "ready for use" in sectors such as health and well-being, entertainment and creative industries, mobility and logistics, public safety and environment and climate. However the gap between the availability of technology and real use must be bridged urgently by joint efforts of research, development, and domain-specific valorization of AIR technology.

...and can be delivered.

The ICT Innovation Platform (IIP) Intelligent Communication seeks to bridge this gap. The outlook for Dutch impact and success in this domain is positive. The Netherlands are forerunner in high-end backbone and wireless/wireline broadband coverage. Experience with public internet is very high because of the open minded attitude towards early adoption of new technology.

4.6.3 Complexity

At present, communication, internet, sensor technologies and applications are integrated in an ad hoc manner. In the coming decades, these domains will merge into an integrated communication fabric. Even more importantly, however, is that this communication fabric will retreat into the background. People will no longer be bound to specific devices such as computers or PDA's. And computational intelligence and communication capabilities will be available and duplicated everywhere. This can be in television sets, navigation systems, mp3 players, sensors in clothes or walls, but also in a variety of sensing systems for instance for health, safety, and environmental monitoring. Compared to the present, the challenge will be much more on effective usage of available sensor networks and communication capabilities rather than on connectivity itself.

4.6.4   Abundance of Intelligent Capabilities

The communication and processing capabilities in the world surrounding us will be abundant and in most cases the different possibilities and choices for technologies, processing and services in the last and first meters of the communication chain will form a highly redundant system. In order to optimally deal with this enormous increase of complexity in the last and first meters we need a systems approach and new intelligent solutions.

4.6.5   The Demand Side

Experience of the past years has shown that innovators in application-specific domains increasingly try to implement new technologies and have a clearer picture what they want to achieve with new communication technologies. In practice, they still encounter many problems with respect to dependability and scalability of wireless solutions, integration with legacy and back office systems, lack user experience of applications, no suitable business models in place and the volumes of data communicated and stored. As a result, many innovations and inventions remain on the shelf, do not find their way into practice, or do not deliver the promised benefits. In addition, the diversity of the application domains and especially the diversity of all the individual applications and communication possibilities require focused attention on application-specific solutions. It is especially this diversity that puts the communications sector for major challenges.

4.6.6   Intelligent Communication Challenges

Summarizing, the challenges lie in finding intelligent communication solutions that

- create domain specific solutions that are optimally suited for the intended purpose, with high user experience and viable business models.

- help us deal with the enormous complexity in the first and last meters of the communication chain
- help the Intelligent Communication sector to meet the demand for highly specific solutions at a reasonable cost.

4.6.7 AIR Concepts

When analyzing the challenges for solutions within the diversity of the complex environments in different domains, there are three common and recurring aspects:

- *Adaptivity*: the overall system is aware of relevant context parameters and adapts to changing context and conditions in a self-aware manner.
- *Intuitiveness*: For the user the system shows the behavior and functionality that the user expects nothing else. This requires careful balancing between autonomy of the system and the control and experience of the user. Intuitiveness also implies that the communication system does more than just moving bits: it connects people with their environment by providing them useful information and pleasant-to-use content.
- *Robustness*: overall system is sufficiently dependable and resistant to undesired changes, which may be occurring involuntarily (noise, errors, connection or power loss) or that may be due malevolent behavior such as security and privacy attacks, and malfunction.

We consider the next stage in communication systems the inclusion of intelligence. The central solution that the IIP Intelligent Communications aims for is to realize intelligence by making the communication environment adaptive, intuitive and robust. This can be achieved by providing both generic and domain-specific communication solutions that optimally utilize the available communication and computation resources.

4.7 Conclusions

Concluding we can state that a thorough analysis of three, relatively arbitrary but important domains in nowadays life shows that a common base of intelligent communication challenges lies at the base of breakthrough innovations in these domains. This commonality is the long-praised stronghold of information and communication technology, and opens up a new era of cyber-physical systems, with necessarily a strong human touch.

References

[1] http://www.eitictlabs.eu/action-lines/

[2] S. Cherry, Edholm's law of bandwidth, IEEE Spectrum, no. 41, pp. 58–60, 2004.

[3] Cisco Visual Networking Index (VNI): Global Mobile Data Traffic Forecast Update, 2012–2017.

[4] R.J. Sluijter, The Development of Speech Coding and the First Standard Coder for Public Mobile Telephony, PhD thesis, TU Eindhoven, 2005.

[5] J.C. Haartsen, The Bluetooth Radio System, IEEE Personal Communications, pp. 28–36, Feb 2000.

[6] W. Lemstra, V. Hayes, and J. Groenewegen (eds), The Innovation Journey of Wi-Fi: The Road Toward Global Success, Cambridge University Press, 2010.

[7] M.K. Smit, C. van Dam, PHASAR-based WDM-devices: Principles, design and applications, IEEE J. Sel. Topics in Quantum Electronics, vol. 2(2), pp. 236–250, June 1996.

[8] http://wirelessgigabitalliance.org/

[9] http://www.tribler.org/trac/wiki/whatIsTribler

[10] European Foresight Monitoring Network, Special issue on healthcare: Healthy ageing and the future of public healthcare systems, Nov 2009.

[11] Mensenzorg. Een transitiebeweging. Rotterdam: DRIFT Transitieprogramma Langdurige Zorg, May 2009. In Dutch.

[12] Semantic web. Accessed at http://en.wikipedia.org/wiki/Semantic_web on 15 May, 2009.

[13] The National Energy Efficiency Initiative, Smart Grid, Smart City: A new direction for a new energy era, Australian Government (department of the Environment, Water, Heritage and the Arts), 2009.

[14] http://www.smartgrids.eu/

[15] http://energy.gov/oe/technology-development/smart-grid

[16] Taskforce Intelligente Netten, Op weg naar intelligentenetten in Nederland, Ministery of Economic Affairs (The Netherlands). In Dutch.

[17] ICT InnovatiePlatform MAIS, Mobiliteit als ICT-systeem, Strategic Research Agenda, 2009. In Dutch.

[18] http://ec.europa.eu/transport/themes/its/road/action_plan/

Biography

Erik Fledderus received his MSc and PhD in the field of Applied Mathematics. Starting in 1998 he worked with KPN Research, the research lab of the Dutch incumbent telecom operator. His main field was mobile networks, including propagation channel modeling, multi-user detection in spread spectrum systems, and UMTS network modeling. He was coordinator and principal architect of the European project Momentum (http://momentum. zib.de/). In addition, he was project leader of the Dutch project Broadband Radio@Hand, on MIMO and radio over fiber — this project has put forward the basics for the 802.11n standard and demonstrated as the first in the world a 3×3 system based on 802.11a (=> 162 Mb/s).

Erik started in 2003 as part-time professor at Eindhoven University of Technology in the field of wireless communication networks. Also in that year, KPN Research was sold to TNO, a leading research and technology organization in the Netherlands. Since 2011, Erik's interests have moved to cognitive principles for radio networks. In addition, he is now managing director of TNO's ICT activities, both business-wise and research-wise. He frequently audits European projects, and acts as advisor to NGMN (Next Generation Mobile Networks).

CV Henk Eertink received his MSc and PhD in the area of Computer Science. He started working at the Ministry of Defense, in real-time software development for the artillery. In 1994 he finished a PhD on formal methods in computer science, and after that joined Telematica Instituut. Henk is currently focusing on service architectures, multimedia, and context-aware systems.

Henk has (co-)supervised 5 PhD students, and submitted 3 first-ranked FP7 proposals as prime author and editor.

CV Patrick Essers after his study Electrical Engineering at the Technical University Eindhoven, Patrick Essers started in 1992 working for the Dutch PTT Telecom in the area of Data communication.

In 1996 he moved on to Ericsson Telecommunication in the Netherlands, where he became specialist in the field of mobile communication.

During his 17th year career he took on several management positions, with his most recent job as Program Director Seedlinqs, the ecosystem for open innovation in the area of multimedia applications.

5

MIMO Systems and Application to Brain Computer Interface by Using EEG

Silvano Pupolin

Department of Information Engineering, University of Padua, Italy

5.1 Introduction

Brain Computer Interface (BCI) is, as proposed by Vidal in 1973 [1], *a multitechnology discipline able to use signals generated by the brain to dialogue with intelligent devices in order to support the person in controlling external apparatus as e.g.,: prosthetic devices.* Also, it could permit to have a better knowledge of neurophysiological phenomena that govern the production and the control of observable neuroelectric events. The capture of brain signals could be done in invasive or non-invasive techniques. The first insert into the brain appropriated electrodes while, the second, capture the signals from the scalp and/or the body of the person. Using both techniques the signals we could capture are either the electric voltage or the intensity of a magnetic field. The simplest way to capture brain signals is to measure the electric voltage between one electrode placed on the scalp and a reference point, by the so called Electroencephalogram (EEG). The signals we are revealing are sustained by fluctuation of either electrical potential or current generated by neurons. An alternative way to detect the same signals is to measure the

COmmunications- NAvigation-SENsing-SErvices (CONASENSE), 99–114.

magnetic field generated by currents in the brain just below the skull, with a magnetic field detector outside the scalp by the so called Magneto Encephalo-gram (MEG). EEG is more robust than MEG to the background electromagnetic field where we are every day immerse, so that is the main used system to detect brain activities to day and several studies have been done during the last decades to design sophisticated machines to interpret EEG signals.

The use of EEG signals to detect what the human brain is doing was first proposed by Vidal in 1973 [1]. The Vidal research proposal generates a lot of study efforts to understand how the human brain works and how to use the information we can derive from EEG. Looking at EEG we note that when the person is in a relaxed state there is an almost periodic signal due to the spontaneous brain activities. If the person is subject to sensory messages, such as sounds, light, etc., this periodic signal almost disappear and is substituted by short waveforms that are evoked from the sensory messages. In order to detect these signals, which have a duration of 0,5–2 seconds, we need to know other information like: the area of brain where they have been generated; the standard parametered signal shapes, if any, the typical signal bandwidth, etc. Many papers treated in the past these topics but they have not reach a fully complete result. For example many papers try to get the position inside the brain of the point where the evoked potential has been generated. In order to solve the, so called, inverse problem, first the electric field on the skull, generated by either a dipole or a current source in a known position within the brain, has been computed. In doing this computation the brain has been considered as a homogeneous material with known and linear electric characteristics. In the analysis two discontinuities made by two layers of homogeneous materials have been considered. The first one represents the cerebral cortex, which appears as a good conductor compared to the brain and, the second is the skull which appears as a good insulator compared to the brain. The result is that the electric field measured on the scalp is deterministically related to the position and the intensity of the source. To solve the inverse problem the knowledge of the electric field all over the scalp is required. Also, we remark that to the received signals generated by the brain overlap noise and interference.

To solve the inverse problem is a challenging engineering task, but, is this the problem we want to solve? Specific points of the brain devoted to perform specific tasks depend on several factors; most of them are still unknown. What is known is that a given task is done in a volume of the brain, while the specific point where it is done depends on the organization of the brain of the person considered. Then, in many cases we are interested to know the area where the signal originated not to the specific point.

In the following we limit our analysis to EEG signals. As next step in our research we introduce new ideas in order to apply to EEG the results of MIMO and radar technologies for a further improvement in the analysis of EEG signals.

The Chapter is organized as follows. Section 2 is devoted to analyse the signal generated by EEG and to classify them. Section 3 poses the new problem of the signal propagation within the brain and introduces the basis for a new model for the brain electrical field propagation based on measures. Section 4 considers MIMO and radar systems for detecting the signals generated by the brain, classify them and localize their origins. Section 5 concludes the chapter.

5.2 EEG Signals and Their Classification

The human brain is the central organ controlling the nervous system and coordinating all human functions (movements, heartbeat, blood pressure, etc.). It is composed by more 100 billion neurons, each connected to as many as 10,000 other neurons. Neurons are electrically active, process the information and exchange signals with other neurons through the *axons* and *dendrites*. From a communications point of view each neuron has a broadcast transmission by using the only axon departing from it and, receives information in a multi-user shape from many other neurons through the dendrites. Each axon, which has a length varying from a fraction of mm to a meter or more, could be *connected* to thousands of dendrites through *synapses*. In physiology the neurons are called *grey matter*, while the axons are the *white matter*. The grey matter of the outer layer of the brain is called *cerebral cortex*. Moreover, cerebrum has two hemispheres (left and right) and each hemisphere is further divided in four lobes which are devoted to specific functions.

Frontal lobes, located in front of the hemispheres, are involved in tasks as language production, working memory, motor function, problem solving, socialization and assists in planning, coordinating, controlling and execute behaviours.

Parietal lobes, located just behind the frontal lobes are involved in integrating sensory information from the body and in the manipulation of objects. One function is the visio-spatial processing.

Temporal lobes, located at the sides of the hemispheres are involved in auditory processing and in semantics of both speech and vision. They contain the hippocampus which controls the formation of new memories.

Occipital lobes, located in the rear part of hemispheres, are devoted to visual processing.

The *cerebral cortex* is the grey tissue covering the surface of each lobe and is responsible for language, music, calculation, imagining, thinking, etc. and controls any part of the body we can move deliberately. The cortex is divided in 52 zones, called Broadman areas [2, 3].

The whole brain is contained in the skull which has a thickness of about 2 mm.

The grey matter of the brain generates electric signals which propagate through the grey and white matter and the skull to create an electric field on the scalp (external part of the skull). We could measure the voltages via one or more electrodes placed on the scalp and a common reference point. These voltages are proportional to the intensity of the electric field generated by the neurons attenuated during their propagation through the brain and the skull (see Section 5.3). The EEG voltages we are measuring are a time varying signals and their characteristics depend on the type of activity the brain is doing. We remark that signal produced by a single neuron is not measurable on the scalp. To get a measurable signal many neurons in the same brain region are activated synchronously, with behaviour similar to the laser effect in electro-optics.

The EEG we measure is typically due to several signals produced simultaneously in different parts of the brain. Because the EEG is composed of several signals detected in different parts of the scalp, what we get is a *Multiple Input Multiple Output* (MIMO) system, where the inputs are the signals generated by the brain and the outputs are the voltages collected from the electrodes positioned in the scalp.

A typical analysis of the EEG is based on its frequency components since it present rhythmical patterns. EEG waves have been classified according to different brain functions, but the terminology is imprecise and sometimes abused because traditionally brain waves were classified on the basis of visual inspection and not using precise frequency analysis. Keeping in mind that there is no precise agreement on the frequency ranges for each type, five main types of EEG waves have been defined.

1. <u>*Delta:*</u> it has a spectrum in the lowest frequency range, below 4 Hz. It is typical for infants and is present in deep sleep and in case of some organic brain diseases.
2. <u>*Theta:*</u> it has a spectrum in the frequency range from 4 to 8 Hz and is associated with drowsiness, childhood, adolescence and young adulthood. This EEG frequency can sometimes be produced by hyperventilation. Theta waves can be seen during hypnagogic states such as trances, hypnosis, deep day dreams, lucid dreaming and light sleep and the preconscious state just upon waking, and just before falling asleep.
3. <u>*Alpha:*</u> it has a spectrum in the frequency range from 8 to 12 Hz. It is characteristic of a relaxed, alert state of consciousness. For alpha rhythms to arise, usually the eyes need to be closed. Alpha waves attenuates with drowsiness and open eyes, and typically come from the occipital (visual) cortex. An alpha-like normal variant called <u>*Mu*</u> is sometimes seen over the motor cortex (central scalp) and attenuates with movements, or already with the intention to move.
4. <u>*Beta:*</u> it has a spectrum in the frequency range from 12 to 30 Hz. A low amplitude of beta waves with multiple and varying frequencies is often associated with active, busy or anxious thinking and active concentration. Rhythmic beta activity with a dominant set of frequencies is associated with various pathologies and drug effects.
5. <u>*Gamma:*</u> it has a spectrum in the frequency range from approximately 30 to 100 Hz. Gamma rhythms maybe involved in higher mental activity, including perception, problem solving, fear, and consciousness.

When a person is in a relaxed state the Alpha signal is typically present. It has a periodic shape which almost disappears (*desynchronizes*) when either a movement activity or one or more stimuli activate the sensory system (e.g., visual or hearing). Thinking is another way to desynchronize Alpha signal.

Remark. If the signal is generated by as many as 10,000 neurons, in case of synchronous signals the amplitude is 10,000 times that of a single neuron, while in case of de-synchronization the average amplitude is proportional to the square root of the number of neurons involved, in this example 100 times. Then when neurons de-synchronize we get an amplitude reduction of the signal, for the example considered, by a factor of 100 with respect to the synchronized case.

Then, if we apply a stimulus to a person in relaxed state we note the de-synchronization of the Alpha signal and the appearance of a new signal correlated with the stimulus, called *evoked potential*. By experiments, if we generate a short stimulus (less than 100 ms long) the evoked potential has a duration ranging from 0.5 to 2 s. The EEG signals we measure exhibit correlations because they are generated by the same brain sources which had different link attenuations. Some brain signals belongs to different classes of brain waves, so a first way to separate out these signals is to consider the correlations among the signals filtered on bandwidths of brain waves. Note that the electrodes measure also noise and electric disturbances due to other causes than the neurons electric activities. One of the purposes of the detector is to reduce the effect of noise and disturbances in order to improve the *Signal-to-noise ratio* (SNR), which is a measure of i.e., quality of the EEG, the bigger is the SNR the better is the EEG. Also, because of the presence of several signals generated by the brain, having many signals detected we could use *Multi-User Detection* (MUD) techniques to separate the signals and a positioning technique based on the amplitudes of the received signals to identify the area where they have been originated.

5.3 Electric Field in the Brain and the Propagation Model

The papers that treated the propagation model in the brain used deterministic studies based on Maxwell's Equations (4). As remarked in (4, Ch. 4), the bandwidth of electromagnetic waves in the brain is within the range 0–500 Hz. The solution of the Maxwell's equations shows that the electric and magnetic fields are almost independent (from a practical point of view they are considered independent from each other), so that electric (magnetic) field could be computed with the classical theory of quasi-stationary fields. Even with this simplified analysis the electric field in the brain cannot be easily computed

Table 5.1. Resistivity of brain material.

Material	Resistivity (Ω cm)
Copper	$2 \cdot 10^{-6}$
Seawater	20
Cerebro Spinal Fluid	64
Blood	150
Spinal Cord (longitudinal)	180
Spinal Cord (trasverse)	1200
Cortex (5 kHz)	230
Cortex (5 Hz)	350
White matter (average)	650
Bone (100 Hz)	8,000–16,000
Pure water	$2 \cdot 10^7$
Active membrane (squid axon)	$2 \cdot 10^7$
Passive membrane (squid axon)	10^9

because of the different constitutes, the brain is composed (grey and white matters, blood, etc.). Each matter has its own electrical characteristic in terms of resistivity, capacity, inductance, and also some of them are anisotropic, i.e., an electrical behaviour change with the direction of the current, as happens in the white matter. As an example we give Table 5.1 in which the values of resistivity of matters present in the brain are reported.

In order to model the electric field on the scalp several authors consider the brain as a homogeneous conductive system. So, by applying either a dipole or a current source in a given position within the brain it is possible to compute accurately the electric field on the scalp and, consequently the voltages we could measure among the EEG electrodes and the reference one. Almost all the models used today are based on the above assumptions and are able to find the area where the signals have been generated. In all cases a simulation run is needed demonstrating that results depend on the shape of the skull.

Based on the experience gained in terrestrial wireless communications it is possible to follow a new direction for channel modelling based on measurements instead of solution of Maxwell's equations. The model we will derive is a statistical one and the statistical parameters should be tuned to the specific patient. This model should take care of the variation of the resistivity of the materials either with the age of the patient or with climate conditions, the presence of anisotropy, moving fluid (blood), etc. In order to collect the data needed to define the new proposed model a common protocol to collect all the data needed should be agreed. We recall that the main difference with respect to terrestrial wireless communications is that a training signal is not available. Then, the best we can do is to put together different diagnostics at the same time, as EEG and Magnetic Resonance (MR), where the last could indicate in a very precise way the area where electric activities surge in the brain. So, we could obtain precise indication of the attenuation vs. distance.

5.3.1   A Preliminary Proposal for Brain Statistical Channel Model

Extracting from what has been stated before we could say that the electric field we measure on the brain at a sufficient distance from where it has been originated has a spherical polarization due to the anisotropy of the resistivity of the brain matter.

The model we are proposing is given by a random variable which has an average value depending on the distance source-electrode and a probability distribution function that should be determined. The average attenuation of the electric field depends on the type of matter it crossed. We could assume that its value in dB be composed by a fixed value due to the attenuation induced to cross the skull plus a value increasing proportionally with the logarithm of the distance. Then the relationship between the electric field $e_g(r)$ generated by the brain at the point r and the voltage $v_s(r')$ measured at point r' in the scalp is given by:

$$v_s(r') = e_g(r)h(r, r') \tag{5.1}$$

with the average value of $h(\bar{h})$ given by:

$$\bar{h}(r, r') = h_0/(|r - r'|)^2 \tag{5.2}$$

where h_0 is a constant value that should be determined.

By considering a spherical polarization of the electric field on the scalp with an angular uniform distribution we obtain a uniform distribution of the amplitude of the electric field orthogonal to the scalp in the range $[0, V_{max}]$. What we are measuring is the voltage between two points in the scalp, which is the integral of the electric field along a line connecting the two points. The values of the electric field along the line are correlated due to the geometry and the physics that generate them. However, the voltage is reasonably represented by a Gaussian random variable whose mean and variance should be evaluated. Also, looking at the position of the electrodes on the scalp as shown

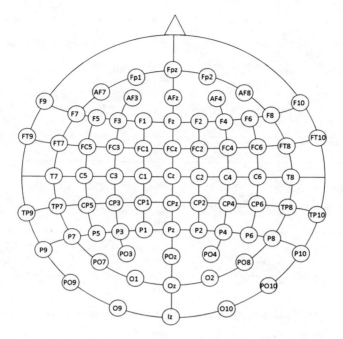

Fig. 5.1 Channel electrodes arrangements for an EEG cap.

in Figure 5.1 for an EEG cap of Brain Products, the manufacturer proposes that the common reference point be the pin denoted by FCz. Some physicians ignore the suggestion and choose the common reference point outside the scalp. The voltages collected by each pin are correlated and the correlations depend on both the pin and the position of the signal source within the brain. Note that in the presence of multiple independent brain signals the correlations between the detected signals from different electrodes are combinations of signal correlations per source.

In Figure 5.2 a sample of a raw EEG signal recorded from a healthy person in a relaxed state is reported. It is possible to note the correlations among the signals and the presence of a pseudo-periodic signal at about 12 Hz, as shown in Figure 5.3 where the signal spectra are reported. Also, a disturbance at 50 Hz with odd harmonics, due to a coupling with the electric power network, is clearly visible.

In order to identify the channel first we cancel the well identifiable interference signals, as e.g., the 50 Hz due to power lines. Then, because the channel

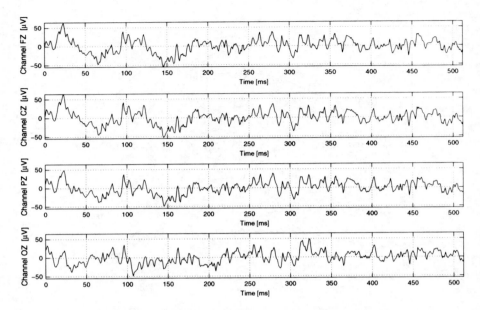

Fig. 5.2 EEG signals derived from pins Fz, Cz, Pz, Oz for a healthy person in relaxed state.

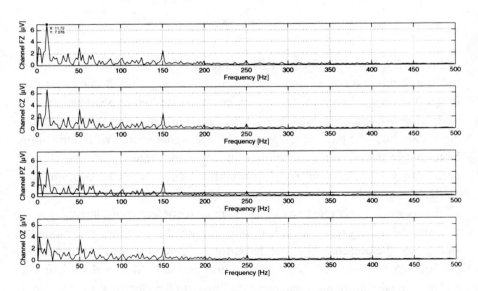

Fig. 5.3 Power spectrum of the signals shown in Figure 5.2.

model could be frequency dependent we separate out the signals generated by different classes of EEG waves by filtering the row EEG signals. Further, for each separable signal we apply multiuser detection in order to get a cleaner signal separation. Once we get the expected transmitted signals we could compute for each of the pins their components and the cross correlation matrices. Then, we could proceed to match the received data with the statistics of the attenuation law, i.e., the law of attenuation vs. distance of the average attenuation, the distribution probability function and its parameters.

5.4 MIMO Techniques to Detect EEG Signals and to Localize Their Origin

The systems we are studying from a MIMO point of view is composed by multiple users (the sources of the brain signals) with a multiple receive antennas. Let N_s be the number of brain signal sources and N_{cap} the number of electrodes in the EEG cap. Then, the input-output relationship is:

$$\mathbf{R}(f) = \mathbf{S}(f)\mathbf{H}(f) + \mathbf{W}(f) + \mathbf{D}(f) \qquad (5.3)$$

$\mathbf{R}(f)$ is the column vector of size N_{cap} reporting the frequency components of the measured EEG data; $\mathbf{S}(f)$ is the column vector of size N_s reporting the frequency components of the EEG source data; $\mathbf{H}(f)$ is the channel transfer function matrix of size $N_s \times N_{cap}$; $\mathbf{W}(f)$ is the vector of size N_{cap} reporting the frequency components of the additive white Gaussian noise (AWGN); and $\mathbf{D}(f)$ is a vector of size N_{cap} containing external electromagnetic disturbances coupled with the EEG electrodes as for example: 50 Hz signals with its harmonics coming from the electricity power network.

5.4.1 Detection of EEG Signals

Because we are in the presence of multi source signals and known external sources of disturbances we could, first, clean-up the received signals from the disturbances. We could track the sine waves generated by external sources in amplitude, frequency and phase and subtract them from the received signals. The best way is to use a MMSE technique to subtract the disturbances. For example, the coupling of electric power sources produces in Europe a periodic

Table 5.2. Amplitudes and phases of the 50 Hz disturbance.

	1st harmonic	3rd harmonic	5th harmonic
FZ			
par Amplitude	6.673	5.266	1.546
Phase (rad)	0.5443	−2.3480	1.2907
CZ			
Amplitude	7.0865	5.407	1.577
Phase (rad)	0.3851	−2.3256	1.406
PZ			
Amplitude	7.425	4.980	1.493
Phase (rad)	0.3086	−2.3124	1.5925
OZ			
Amplitude	7.463	4.770	1.5246
Phase (rad)	0.2784	−2.3283	1.6873

signal at 50 Hz; typically also the third and fifth harmonics are present. This could be seen from Figure 5.3 where the spectrum of each of the four EEG signals shows a spectral line at 50, 150 and 250 Hz.

The values of the amplitudes and phases of the three cosine waves are reported in Table 5.2. The effective frequency detected is 50.06 Hz.

Once we have a cleaned signal we could proceed with the analysis. In some cases, as e.g., when a patient is in the relaxed state, the signal is periodic (Alpha signal), while when the EEG signals are produced by evoked potentials they are a-periodic and span for a period up to 2s. The shape and the frequency range of such signals are related to the specific stimulus that generate them and to the general physical state of the person. In the case of Alpha signals we can proceed to identify the sine wave and all its harmonics by using the same MMSE algorithm used to cancel the interference, where we should also estimate the frequency of the sine wave. In the case of evoked potentials we have to estimate unknown continuous time signals. The way we foresee to identify the specific stimulus is the use of radar techniques to classify all the possible signals generated by evoked potentials. Because of the presence of multiple electrodes and multiple signals at the same time the best we can do is to use Multiuser and Multisensor detectors. Note that in radar communication the system transmits a known bursty signal and receive the same signal delayed and distorted by the reflection on the target and all other objects present in the scene. The target has its own signature which is known and stored in a data base. Then, a detection technique is used to match the received signal shapes with one of those stored as reference. With respect to the stored signal shape the one received is further distorted due to a different angle of view of the target and of its speed.

We need to proceed in the same way also for the classification of the EEG signal. First, it is necessary to collect many evoked potentials EEG shapes. Then, on the basis of the detected signal we match it with one of the stored one. The problem we need to solve is related to the fact that for a given stimulus the EEG signal shapes are different from person to person and, for the same person it changes on the basis of physical condition, as: tiredness, illness, etc. What we need is to prepare the data base and later to test the classification algorithm which uses it. Also, there are different operating conditions, which require different sets of data bases. Two examples are presented here in order to show what is needed to do.

Example 1. We are looking to a laboratory which is in charge to check if an EEG is from a person in good health or it is moving toward some disease. It is important to have a screening for illness at its beginning, so it is important that the system which is evaluating the EEG is able to identify any possible deviation from standard values. The system is checking many different patients and need to be able to track the main functionalities from EEG from a standard set of tests. The data base behind the system is large and the analysis of EEG could require long time for processing all available signals. It could be done on computers which have no problems related to energy consumptions.

Example 2. It regards an EEG connected to BCI in order to respond to the requirements of a disabled person. The system for most of the time is battery powered, so energy consumption is a key parameter. Also, the system could be tailored to the person to whom it is connected and should give a response to the EEG detected stimulus in short time, equivalent to those of human reaction. The data base is simpler than the one of Example 1, but one needs fast decisions.

5.4.2 Positioning of EEG Sources into the Brain

In Paragraph 5.4.1 we set-up algorithms able to identify the evoked potential event. Now, on the base of the received data we need to identify the origin, within the brain, of the evoked signal. If we know the law of propagation vs. distance within the brain we could, on the basis of the received energy in each electrode define a triangulation algorithm, similar to the one used for navigation systems, to identify the area where the signal has been generated. The fact that each signal is corrupted by noise and interference and that the propagation model we are looking for is non deterministic, the area where the signal has been generated is given as a probabilistic set. What we obtain is a region in which the detected signal has been generated with a probability higher than a given threshold.

5.5 Conclusions

The detection of EEG signals is a new non-invasive technique to help disabled persons to return to their normal life and to improve rehabilitation techniques. To detect the signals many papers have been published and several books treat the problem. Most of the results were based on the fact that the brain has been considered to be composed by three homogeneous layers, the brain, the cerebral cortex and the skull. This is not true because the material of the brain is neither homogeneous nor isotropic. So, the proposal of a measurement campaign to define a model for the electric field propagation in the brain is highly recommended.

Also, the use of radar techniques to classify the signals induced by evoked potential is a new paradigm. It requires a long measurement campaign to identify most of the EEG signals and their modifications due to illness, aging, etc., and to connect them to the proper stimulus. The creation of a data base that could be used worldwide will give a great benefit for prevention of illness due to nervous diseases, as Parkinson, Lateral Amyotrophic Sclerosis, etc. The necessity of identifying signals and to evaluate their correlations in order to avoid erroneous classification and identification of diseases is a must of future research.

References

[1] Vidal, J. J., "Toward Direct Brain-Computer Communications," *Annu. Rev. Biophys. Bioeng.* 1973, Vol. 2, No. 1, pp. 157–180.

[2] Broadman, K., *Vergleichende Lokalisationslehre der Grosshirnrinde in ihren Prinzipien dargestellt auf Grund des Zellenbaues.* Leipzig: Johann Ambrosius Barth Verlag, 1909.

[3] Garey, L. J., *Broadmann's Localization in the Cerebral Cortex*, New York, NY: Springer, 2006.

[4] Nunez, P. L. and R. Srinivasan, *Electric Field of the Brain.* 2nd. New York, NY: Oxford University Press, 2006.

Biography

Silvano Pupolin, received the Laurea degree in Electronic Engineering from the University of Padua, Padova, Italy, in 1970. Since then he joined the Department of Information Engineering, University of Padua, where he is Full Professor of Electrical Communications. He was:

- Chairman of the Faculty of Electronic Engineering (1990–1994),
- Chairman of the PhD Course in Electronics and Telecommunications Engineering (1991–1997), (2003–2004)
- Director of the PhD School in Information Engineering (2004–2007),
- Chairman of the board of the Directors of the PhD Schools of the University of Padua (2005–2007)
- Member of the programming and development committee of the University of Padua (1997–2002),
- Member of Scientific Committee of the University of Padua (1996–2001),
- Member of the budget Committee of the Faculty of Engineering of the University of Padua (2003–2009)
- Member of the Board of Governor of CNIT "Italian National Interuniversity Consortium for Telecommunications" (1996–1999), (2004–2007)
- Director of CNIT (2008–2010)
- Director Dept. Quantum and Radio Communications of CNVR (Consorzio Veneto di Ricerca) (2011–)
- General Chair of the 9-th, 10-th and 18-th Tyrrhenian International Workshop on Digital Communications devoted to "Broadband Wireless Communications", "Multimedia Communications", and "Wireless Communications", respectively;

- General Chair of the International Symposium "Wireless Personal Multimedia Communications (WPMC'04)" Abano Terme, Padova, Italy, September 2004.

He spent the summer 1985 at AT&T Bell Laboratories on leave from Padova, doing research on digital radio systems.

He is actively engaged in researches on broadband mobile communication systems, personal communication systems, MIMO systems and applications.

6

Multimedia and Network Quality of Service

Oleg Asenov[1] and Vladimir Poulkov[2]

[1] *Department of Mathematics and Informatics, University of Veliko Tarnovo, Bulgaria*
[2] *Faculty of Telecommunicaitons, Technical University of Sofia, Bulgaria*

6.1 Introduction

The term multimedia contains connotation connected to the result of combining various types of media with the aim of presenting more comprehensively properties and features of objects, phenomena, processes and events, etc. Information is the point of intersection between multimedia and information technologies [1]. Computer based technologies' potential to combine and reproduce synchronously diverse media in real time provides multimedia with so far unsuspected possibilities for integrated multimedia presentation of information [2, 3]. Despite the fact that multimedia comes into being millennia ahead of computer and communication technologies, the introduction of the latter brings about a revolution in the use of the multimedia approach for presentation of practically all kinds of information types, ranges, volumes, and contents

Another revolutionary change that contemporary information and communication technologies provide to multimedia development is the possibility of performing interactive multimedia [4]. Due to interactive multimedia the

COmmunications- NAvigation-SENsing-SErvices (CONASENSE), 115–141.

recipient exerts influence over (is provided with the possibility of controlling) the real time multimedia impact and is able to suppress a certain type of media and to modulate the active presentation by another type of media [2]. These possibilities are designed and introduced in advance when developing the so-called interactive multimedia scenario [5]. This innovative approach solves to a great extent the main contradiction accompanying the development of multimedia information presentation technologies — the one between the multimedia information impact potential and the subjective multimodal perception profile. Interactive multimedia development provides for a greater effectiveness when information is presented and hence for a greater authenticity of perception. The visible part of the multimedia iceberg is impressive and effective but there is a complex innovative architecture of computer and communication technologies underneath [3, 6]. A new cognitive term is formed naturally: multimedia networks.

6.2 Differentiating the Networks. Development Processes, Merits and Sublayers

The necessity of differentiating such class of networks arises with the development of the interactive multi-user multimedia which increases the requirements to communication infrastructure. Substantially is extended the range, structure and connections between protocols supporting multipoint multimedia links at session layer [7] of the Reference model [8, 9]. The architecture at this layer is decomposed into several sub-layers which have the task of providing transport of every type of multimedia on the one hand and of synchronously delivering the integrated information volume of every type of multimedia on the other, so that the multimedia presentation is performed and played back in real time [10]. When a multimedia presentation is played back off line no substantial attention is paid to the problems connected to the bandwidth, as far as playback quality depends on computer system configuration only. If system configuration parameters are higher than those of the specific multimedia presentation then the multimedia information presentation set as per scenario shall achieve the expected functionality and impact.

This is not the case with a network-based multimedia application. If a part of the information sources is remote from the target computer, on which playback is performed, then *delivery* of the required volume of digitally compressed media information is of crucial importance to the multimedia impact quality. This problem becomes even more complicated in case that this presentation should be played back synchronously on several target computers which are separated at 'continental' distance from each other. This is an additional argument in favor of differentiating network multimedia and of the necessity of its studying as specific applied network architecture [1].

6.3 Multimedia Networks and Various Media Types

From characteristic fields' point of view various examples can be given in which network multimedia specific peculiarities are manifested establishing it as a separate class of applied network architecture [8]. Multimedia networks can be defined as an aggregate of device and programming means providing the possibility of combining various types of media, regardless of the place it is entered and stored for the intentional information presentation on one or more interconnected consumer computers. In this case quality of transport of multimedia data on the route between source and receiver is a substantial applied characteristic for computer multimedia network consumers.

Figure 6.1 shows a general classification of media types and subtypes depending on the effect of communication characteristics of the transmission medium on the overall media presentation of the information — bandwidth, time delays, communication errors and necessity of servicing real time interactive sessions and network applications [2].

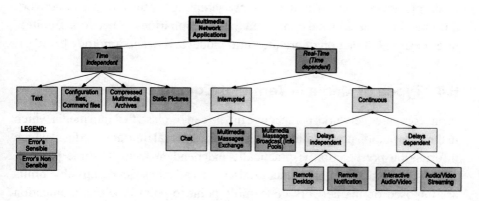

Fig. 6.1 Media types classifications.

Text as a group does not include just the usual series of symbols of the respective alphabet but also various control and formatting symbols, mathematical expressions, as well as symbols for alternative presentations and designations: speech transcription, musical text — notes writing and other modern forms of symbolic presentation of information like for example hypertext.

As long as digital data is transmitted in computer networks the issue of digital presentation of each type of media is essential to the multimedia traffic evaluation. Text and text group media are historically presented and transmitted by a finite number of logical conditions — i.e., in digital form. Morse code is a typical example of logical text presentation — two logical levels in the transmission medium: '.' (dot) and '–' (dash), or in other words a discreet duration (time length) modulated sound. These levels are sufficient for presentation and transmission of characters, digits, and special symbols with a practically unlimited volume. Visual and sound media groups are of a typical analogue nature and in order to be transmitted over the communication channel conversion from analogue to digital way of presentation is performed for these media before transmission. This conversion is generally designated by the term sampling.

The subject of approaches, methods, technologies and algorithms for conversion of analogue media into digital is studied in [11]. Additional conditions and restrictions concerning sampling mechanisms are provided for network multimedia. They are connected to the fact that simultaneously and synchronously a group of media has to be transmitted over the communication channel with bandwidth and time delays in the network data transmission medium known in advance. In this sense sampling for the purpose of network multimedia is specific and connected to the peculiarities of network architectures design, interaction of architectural model layers and topology [9, 12].

6.4   Types of Media in Terms of Computer Networks

From computer networks perspective media can be classified into media which in the information presentation process develop in real time and media in which information presentation is practically independent from the time parameter [10]. For real time type of media strict or relatively strict permissible limits are imposed on the network concerning point-to-point type communication quality key parameters like packets delay (delay) or the distance variation in time between two consecutive packets (jitter). For the time independent type of media like text or image delays or distance in time between two consecutive packets transferring the media file are not critical but bit errors or the Bit Error Ratio (BER) are of substantial importance as they can bring about a substantial change in the content of information presented.

Two basic approaches to authentic data transmission control are known for computer networks in general and computer multimedia networks in particular [6, 13]:

- Automatic request for retransmission in case of an error approach (Automatic Repeat Request — ARQ Request)
- Continuous control and error correction during transmission approach.

With the automatic request for retransmission in case of an error approach the data packets transmitted to the target network address (computer) are controlled continuously and when an error is discovered a service packet is sent to the transmitting computer containing the identification number of the corrupted packet. Since other packets are transmitted over the communication channel after the corrupted packet until the error is detected, the problem with retransmission appears. If the transmitting computer transmits again only the corrupted packet then there will be an inconsistent exceptional packet in the sequence of packets sent to the receiver. For this reason the automatic request for retransmission approach is developed further and it is applied by means of two techniques for corrupted packets retransmission:

- Corrupted packet selective retransmission (selective reject) technique. With this technique the receiver supports a temporary buffer for messages incompletely received, which wait for the retransmission of the corrupted packet.
- Return back to the corrupted packet (Go-back-N) technique. With this technique when an error is detected the receiver rejects all packets received after the corrupted packet. The transmitting computer retransmits the corrupted packet as well as all packets following it regardless of whether these have been received correctly or not.

A communication between two computers in which transmission errors and packets succession is controlled by the automatic retransmission request approach is called connection oriented communication. This type of communication is supported by one of the most often used network protocols TCP (Transport Control Protocol), part of the TCP/IP protocol stack. Automatic retransmission request approach application provides connection oriented communication servicing of applications. From multimedia traffic point of view connection oriented network servicing is applicable to the *time independent* (TI) media type. TCP/IP protocol stack provides error free data transmission with the transmitting computer receiving confirmation from the receiving one [11]. With this type of network protocols the transmission time depends on communication channel quality. When data transmission is accompanied by a lot of errors the transmission time may increase substantially. With connection oriented protocol stacks it cannot be guaranteed that a data unit will be transmitted over the transmission channel within one and the same period of time or within a sustainable time delay interval. This makes the ARQ Request approach inapplicable to the *real time* (RT) type of media.

Requests for servicing of RT media type in network architecture are provided by non-connection oriented protocol stacks like UDP/IP. The UDP protocol (User Diagram Protocol) in IP based network architectures transmits packets from transmitting to receiving computer without a confirmation of an error free packet receipt being necessary [14]. Additional information (packet wrapping) is added to the user data in the packet. This additional information is called *service data for control and correction of errors during transmission*. Specialized algorithms for control and correction of bit errors eliminate the necessity of retransmission as long as a substantial part of the errors due to interferences in the communication channel are corrected by the receiving computer. Data is transmitted with a predetermined error probability, which is sufficiently low to not affect the media presentation quality.

It is substantial with this approach that packets are transmitted with a fixed delay within a certain interval and that this type of protocols provides a minimum time interval before the next data packet is received and processed. This parameter is very important when voice or video communication is transmitted in the computer network, interactive voice or video communication is carried out or synchronized data for presenting impacts on other type of modalities are transmitted in real time (motion sensors, biometric data, etc.) [12, 15].

Transmission time losses are substantially reduced by means of error control and correction techniques embedded in the protocol stacks. And while 5 to 7 years ago the application of these techniques was limited by the relatively lower productivity of the processors embedded in the communication equipment, now with the current development of microelectronics and the relatively more complicated, in terms of computation, algorithms in the protocols without confirmation are processed substantially faster. Error correction without repeated packets transmission is known as FEC (Forward Error Correction) technique — one way error correction. FEC application in network based multimedia applications is effective for RT media type as well as for TI media type while it is only necessary that the estimated BER probability of in the data stream will not aggravate presentation quality and will not affect the authenticity of data presented.

6.5   Discrete and Continuous RT Media

From communication point of view a secondary classification by subtype has to be introduced concerning RT media: discrete RT media (DRT) and continuous RT media (CRT). The classification indicator here is whether in order to service the media presentation is carried out a real time transmission of independent information units — files or messages (DRT) or of a continuous stream of packets interconnected in a succession or in another functional linkage (CRT). DRT media are applied on a mass scale in text oriented interactive applications for communication and data exchange over the Internet [16] and Intranet [17] (corporative, department, internal company networks) like for example MSN/Yahoo Messenger™ or Skype™ chat session. Transmission errors are substantial for DRT and in addition to the standard one-way control and correction algorithms for maintenance of such media network applications, secondary authenticity control in the so called *high* layers of the ISO Standard model [9] and most of all in the *presentation* layer are used.

In this respect one more layer of classification can be introduced — delay dependent and delay independent DRT. An example of delay independent DRT is the information pools which transmit one-way over the Internet to a limited circle of subscribers market information, currency quotations information, meteorological information, road traffic information, etc [14]. With this type of media loss or error during transmission of a message about the current status of the information presented object or process does not lead to a substantial problem with the authenticity of the information because each message with an error is rejected by the receiver and the next message is expected as information pools usually broadcast current information every 10 to 30 seconds.

Impact of delays in the communication channel, on the other hand, is important to the real time network multimedia. In this sense delay sensitive and independent RT type media can be differentiated. Delay sensitive is audio and video streams generated by interactive communication sessions — IP telephony, audio- and videoconferencing or remote desktop or virtual machine access sessions. Delay independent is real time or upon request IP-radio and television broadcasting. With a well-chosen buffer size at the receiving computer the ongoing changes of bandwidth and packets effective delay can be compensated in time [2].

In this sense, when designing a multimedia network application provision of a possibility for adjustment of the applicable, apprehensible to the user traffic limitations is important. This should be made through the estimation of the network connection type (to what extent is a connection pliable to errors, as for example wireless networks are expected to be more pliable to errors from interferences in the air compared to the wired ones, etc.) and the type of media or media operated, as well as media's sensitiveness to traffic limitations. It is important to mention that traffic limitations are relatively independent from each other because, for example, bandwidth determines the number of packets that can be transferred per unit time and this parameter affects, although indirectly, the interval between two successive packets.

6.6   Functional Limitations Related to the Integration of Multimedia Applications

Functional limitations are associated with the integration of multimedia applications in the architecture of the network connections and with the mechanism of interaction between the connected computer systems. When maintaining network connections in the architectural aspect are accomplished, three basic types interaction profiles of computer systems are distinguished [18, 19]:

— Point-to-point connection (unicast).
— Point-to-many points connection (multicast).
— Point-to-all points connection (broadcast).

One of the possible approaches to improving the efficiency of multimedia traffic transport is related to 'exporting' part of the control functions to the session layer of the Reference model [9]. The main idea of this approach is that traffic limitations are to be satisfied at the level of establishing the network connection, in the routing logic being introduced by an additional expanded marking or description of the expected series of network diagrams, which carry a sensitive to traffic limitations multimedia data stream. Such an approach is only possible if all routers in the configuration support QoS functionality, distinguish different priority traffic streams and provide for these streams routes along which traffic limitations are satisfied [20]. Such a configuration can be guaranteed over Intranet and is more difficult to be guaranteed by the Internet suppliers since they provide the traffic limitations in the global network access point and the mechanism of user diagrams transport remain transparent to the user, i.e., independent from the priorities set (marked in the packets) by the user applications [16, 17]. Following phases can be defined in the approach with 'sessions' quality control of multimedia streams network servicing [7, 21]:

- **Phase 1. Media session description:** This functionality provides allocated multimedia applications with the option to exchange and broadcast session information (identification) on media type (audio, video or other data [15]). This is information exchanged in the running session, in the mechanism or algorithm adopted for information compression (JPEG, MPEG4), at session start and end, for description of IP addresses of multicast servers or other servicing computers that provide the session, as well as other applied session information [19]. Based on the description of the media session the resources required for each user for joining in can be associated depending on the set of services requested and the individual user profile for access to the session, as well as in view of the factual limitations which the network connection of the specific user imposes in respect of the access to the multimedia session resources described.
- **Phase 2. Publishing the multimedia session to the potential participants.** This type of description presents the potential development of the multimedia session in time and the potential scenarios for participation of the users to the multimedia information

broadcasted within the specific session. It is important to note here that multimedia sessions may develop according to static scenarios in time, i.e., according to a predetermined program. Access sessions to Internet based radio and television are of this type. On the one hand, every user may plan in advance which transmissions to subscribe and to create his individual profile for access to a set of transmissions on various channels, a program published in advance being taken account of. On the other hand, provided the session is announced on demand, the user creates his individual access profile as a series of requests for the provision of multimedia services with certain characteristics.

- **Phase 3. Media identification in the multimedia session.** Every multimedia session connection is characterized by the possibility of information to be presented by more than one type of media [7]. This requires the supporting of many active media streams — continuous (audio and video) and discrete (visuals and text) which have to be identified. Transmission of audio, video and discrete media in independent data streams provides the users with lower connectivity speeds with the possibility of choosing a limited set of media for media presentation while this does not change the character of the session connection on the part of the transmitter. In case we have users with high quality connectivity, they have the possibility to request all media streams of the current session. In this way, through identification and relatively independent distribution of the media in distinct datagrams the streams are supported by connections which are flexibly adjusted to the changes of bandwidth and network servicing quality.

- **Phase 4. Multimedia session management.** The internal connection between information streams presenting each media is specific for multimedia sessions' management. This requires introduction of intra-media synchronization mechanisms by putting in time markers. For example, in a multimedia session for a virtual visit of a football match the picture from the field is media *video*, the stadium audio background is media *audio 1*, the match commentator is media *audio 2*, teams, football players and coaches statistics is discrete media *on request*. If there are no markers for synchronization between video, audio1, audio 2 (commentator is behind the on field situations), then the network multimedia session virtual ticket for a football match will not provide the expected feeling of the user that he is a virtual real time spectator.

6.7 Internet Architecture Adaptation to Distributed Media Applications and Phases of Time Delay Formation of Multimedia Packets Over the Internet

A typical medium for the functioning of the distributed multimedia applications is the Internet global network. As a network architecture, Internet has been developed with the aim of providing data transfer or transmission. Internet's architecture is not oriented towards a specific end-user and their specific requirements. Internet's basic architecture is not designed to service multimedia traffic. Provision of Internet servicing of distributed multimedia applications will require development and application of additional techniques and approaches so that the Internet architecture is adapted to the graphical and functional limitations of the distributed multimedia applications.

In Figure 6.2 the time setting phases of multimedia packets *video* distribution [15] over Internet architecture are shown.

6.7.1 Generating Multimedia Packets

This phase requires time for conversion of the original media into digital, packaging of digital data into packets compatible for transport over the Internet, addressing, numbering and marking of packets for the purpose of efficient transport over the Internet architecture and reaching target user of the distributed multimedia service. The duration of this phase depends on the media type, correction method and productivity of the producing computer configuration.

6.7.2 Local Distribution of Packets

At this phase packets are transported to and from Internet access points in the local network configurations of the source and receiver. Here on basis of MAC transport in the local network the multimedia packets are delivered to the Internet access point in the transmitting side and from the access point to target

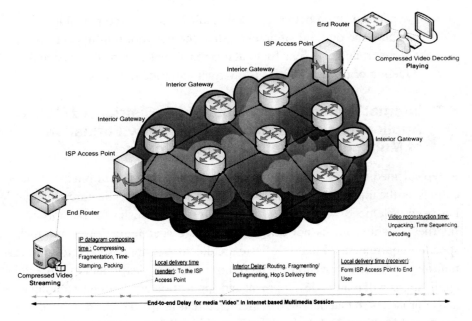

Fig. 6.2   Phases of time delay formation of multimedia packets over the Internet.

user's computer system — in the receiving side. Local networks transmission rates are usually high and MAC protocols generate small time delays. The main problem at this phase is to provide the interaction between MAC protocols and Internet since addressing at the local network level is performed on MAC address basis while distributed multimedia applications over the Internet exchange information by means of IP addresses and IP address logics.

6.7.3   Routing, Buffering and Global Distribution of Multimedia Packets in the Internet

At this phase the substantial time delays of the packets are generated. One of the trivial reasons for that is that packets travel physically long distances along the communication lines and this results in the formation of transport delay. For example, a 20,000 km. optical transport of a packet along Sofia — Santiago De Chile route generates a physical time delay of between 80–100 milliseconds. If we compare this delay with the threshold value of 200 milliseconds for a telephony type service, it becomes clear what a substantial percentage of the total transport delay is the share of the physical delay of the packets along the communication lines. While physical transport delays are determined and practically dependent only on the character of the physical transmission media and on the channel coding method, the delay in the active architectural elements of the Internet (gateways and routers) is determined by a complex of factors. In terms of the theory of mass service the Internet transport structure may be considered as a M/M/1 queuing structure where time delays are formed at the input of the communication lines and in the servicing element itself (gateway or rooter).

Since every communication connection in the Internet internal architecture has a limited bandwidth and is characterized by a final transmission delay, total time delay optimization is only possible when the efficiency of the servicing element of the Markov's network model, i.e., the efficiency of the routing logics through selective prioritizing of the packets, is increased. This improvement of the Internet architecture is known as servicing of packets streams with different level of delay criticality model (Integrated Service Model) and it is provided by a new generation of routing protocols (and active equipment to support them) sensitive to different packets streams — MPLS (Multi-Protocol Label Switching) [22]. MPLS provides fundamentally new architectural possibilities for support of distributed multimedia applications over the Internet. Through marking and identification of streams priority servicing is provided of media data packets that are critical to time delay in relation to those to which variable time delays do not perform a substantial effect on functionality.

6.8 Development of New Models for Servicing of Applied Sessions for Data Transmission in the Internet Architecture

The development of distributed multimedia applications, their extensive popularity among users and the need of providing adequate servicing are the reasons that the development of new models for servicing of applied sessions for data transmission in the Internet architecture is required:

- Integrated network services provision model (Integrated Services — Intserv) [23, 24].
- Differentiated network services model (Differentiated Services — Difserv) [25].

- Multiprotocol label switching model (Multiprotocol Label Switching — MPLS) [22].

Development and application of these models enrich Internet architecture and orient it towards servicing and satisfying traffic limitations of distributed multimedia applications [24, 26]. Joint application of all three models leads to the provision of a new Quality of Service oriented Internet architecture [20, 25]. The term Quality of Service (QoS) is a summarized expression of the requirement claimed for manageability of the Internet architecture concerning servicing of applications with a various levels of criticality regarding the data transmission basic metrical characteristics [20]. To guarantee a certain quality of service a set of metric indicators should necessarily be defined on basis of which this quality is to be assessed and guaranteed. Three metric indicators can be determined for the distributed multimedia applications, which to a great extent characterize and ensure the quality of service of this type of applications in the Internet architecture: bandwidth, delay and reliability.

6.8.1   The Main Drawback of these Models

A major disadvantage of these models is that they regard the multimedia traffic as a quantity of critical data transmitted from source address to target address with a certain quality of service. Due to this approach the link is lost between the specifics of the multimedia traffic as prognostic characteristics and the dynamics of the routing logics reaction to the changes of these prognostic characteristics in time.

6.9   Modern Routing Algorithms in the Internet (IGRP, EIGRP)

The up-to-date routing algorithms in the Internet such as the Interior Gateway Routing Protocol (IGRP) and Enhanced IGRP (EIGRP) are based on the logics of choosing the next node on the target IP address of the datagram in accordance with a routing table previously constructed over one or more metrics, without describing and/or supporting information on the complete route (on the complete network topology respectively, which for the Internet is practically impossible). This procedure applies for every packet arriving at the node (router), the result being the choosing of an exit line over which the datagram of the selected packet is to be broadcasted in accordance with the routing table and the target IP address to the next node [27]. The main problem with this type of routing is that there is no direct relation between the label of belonging to a stream and routing logics. This may lead to a dynamic change in routing and displacement in the sequence of arrival of packets at the target node.

No substantial difficulties will arise if this happens during a file exchange session since packets that are not in series can be buffered to wait for a delayed packet. This alternative is not applicable to multimedia applications and for that purpose IP routing has to be replaced by IP switching. In this case packet marking is used for identification of a logic channel consisting of switches connected in series. A short label is attached to the packet. This label is renewed in each switch based on the information in the label for multimedia stream belonging. Based on the short label the switch makes a decision to which exit port the packet is directed. This switching method substantially speeds up packets directing process and reduces switch overloading.

This approach is well known and it is applied in Asynchronous Transfer Mode (ATM) switches. Correspondence logics are: Virtual Path Identifier (VPI)/ Virtual Circuit Identifier (VCI), i.e., virtual route/virtual circuit of the packets is analogous to the logics applied by Multiprotocol Label Switching (MPLS) [22], the difference here being that MPLS is applied in IP networks. The concept is called multiprotocol since it can be applied integrated with the logics of any protocol of the high layers of the Standard [9], which uses IP network as a transport media. Label based switching has a low impact on apparatus requirements for the active equipment. It provides high productivity of packets directing and ensures flexible traffic management depending of the current congestion on the network architecture [28].

Similarly to the Difserv model [25], MPLS networks also decompose into domains with edge nodes, called Label Edge Routers (LER), and internal nodes, called Label Switching Routers (LSR). A label is affixed by the entry LER for packets entering a MPLS domain and switching inside the domain is performed by checking the label in the switching table. Labels determine the quality of service for the specific data multimedia streams which are distributed in the domain. The label for the next domain is affixed in the exit LER. This is done on basis of a label that is global for the stream and is called Label Switched Path (LSP). The new local label is formed on basis of the Service Level Agreement (SLA) between MPLS domains and the LSP label. Like for

Difserv, the general quality of service requirements for MPLS are also formed on basis of the service request and are indicated in the label [29].

SLA based LSP label transforming between MPLS domains leads to the formation of a managing message to the internal switches in the next domain, which is called Forwarding Equivalence Class (FEC) of service. FEC provides inheritance in every following MPLS domain of the requested quality of service during IP network connection initiation by the multimedia application. MPLS's advantage compared to Difserv is the option of providing a more flexible traffic management through the FEC logics [29]. When changing quality of service the levels of granulation are much more with MPLS compared to Difserv, since switching as a process allows more precise traffic manageability and more effective bandwidth utilization at substantially lower transit delays in the switches.

6.10 Experiments

The following examples illustrate the importance of forecasting and analyzing the properties of the multimedia traffic to be transported and extraction of the values critical for the routing logics such as estimated maximum required bandwidth and estimated maximum size of the user packet for the next slot.

Figure 6.3 shows the testing set up for carrying out experiments for estimation of video scene's basic properties in relation to the key routing parameters — user packet maximum size and maximum generated packets stream intensity, reduced to bandwidth of the communication channel.

The testing set up on Figure 6.3 is developed to demonstrate the functionality of the phase of preliminary analysis and prognostics of key parameters of multimedia traffic to be transported [30]. Various methods and algorithms for dynamic video images compression [15] are investigated in the experiments. For the performance of the tests two standard clips are used, constructed by continuous repetition of a series of frames aiming at obtaining a relatively constant image complexity, measured in %, and of motion, which is also measured in %:

- Clip 1: 80% average image complexity and 80% motion in frames.
- Clip 2: 50% average complexity and 50% motion in frame.

The graph on Figure 6.4 is formed by the test data (A screenshot from the diagnostic programming module CCTV IP Design Tool™). The data is

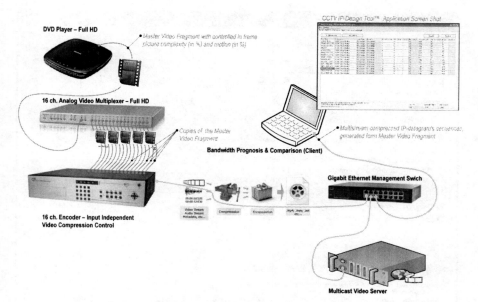

Fig. 6.3 Experimental setup for testing video type multimedia transmission.

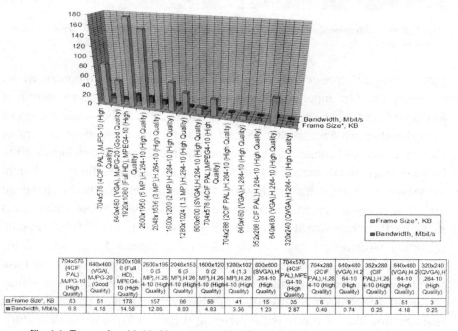

	704x576 (4CIF PAL) MJPG-10 (High Quality)	640x480 (VGA) MJPG-20 (Good Quality)	1920x1080 (Full HD) MPEG4-10 (High Quality)	2600x1950 (5 MP) H.264-10 (High Quality)	2046x1536 (3 MP) H.264-10 (High Quality)	1600x1200 (2 MP) H.264-10 (High Quality)	1280x1024 (1.3 MP) H.264-10 (High Quality)	800x600 (SVGA) H.264-10 (High Quality)	704x576 (4CIF PAL) MPEG4-10 (High Quality)	704x288 (2CIF PAL) H.264-10 (High Quality)	640x480 (VGA) H.264-10 (High Quality)	352x288 (CIF PAL) H.264-10 (High Quality)	640x480 (VGA) H.264-10 (High Quality)	320x240 (QVGA) H.264-10 (High Quality)
Frame Size*, KB	83	51	178	157	96	58	41	15	35	6	9	3	51	3
Bandwidth, Mbit/s	6.6	4.18	14.58	12.86	8.03	4.83	3.36	1.23	2.87	0.49	0.74	0.25	4.18	0.25

Fig. 6.4 Test results with 14 video streams with different resolutions and compression type.

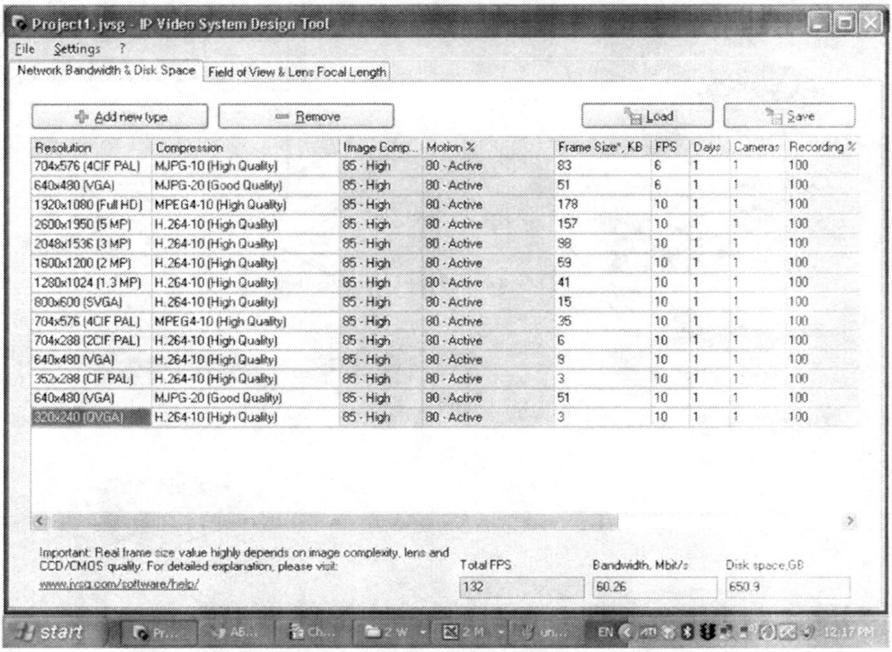

Fig. 6.5 Results with randomly selected methods for compression and resolutions, applied to the reference input video stream.

shot at 'unicast' connections from the Encoder to the test computer, where for every Encoder input a different type of compression and resolution at Standard Clip 1 has been programmed (Figure 6.5). Two basic parameters of the video stream are prognosticated and studied in this experiment, i.e., frame size generated and connection minimum required throughput (bandwidth) to be transmitted and decoded [31] without a loss of the typical for the (adherent to the type of compression) quality. The width of the interval for the change of the frame size is observed, which reflects changes in the network with respect to the fixed packets size at channel and network level [32, 33]. If information regarding prognostic frame size is added as an extended parameter to the MPLS 'label', the packets fragmenting/defragmenting logics would perform with higher efficiency and when choosing a route, additional limitations may be imposed related to the currently prognosticated maximum packet size. In a real situation like the one in the experiment (Figure 6.3) the network medium is supposed to transmit frames with a size of 1 KB to 180 KB, the expected delivery time and distance in time between two consecutive frames in the decoder being one and the same, since frame speed is fixed for every channel to 10 frames per second.

The substantial advantage of maximum frame size and necessary bandwidth estimation can be seen even at this stage of the tests and even with different versions of one and the same compression method — H.264.

Figures 6.6, 6.7 and 6.8 show the results for estimating the key characteristics of different compression methods at a fixed resolution of 1920 × 1080, correct Full HD for Standard Clip 2, for the bandwidth (Figure 6.6) and for the frame size generated (Figure 6.7). MJEPG-1 compression method is used for comparison [34]. It is applied as a standard of highest image quality without losses during compression and decompression. The data and graphs confirm the usefulness of introducing a phase of estimation of the multimedia traffic parameters at session layer. As expected, from experimental point of view, H.264 is substantially superior to all other compression methods. The most important, advantage of this method from communication point of view is the small frame size — 20 to 30 KB for a Full HD frame. The number of fragments when transmitting a frame is substantially reduced, hence service information volume is also reduced and the bandwidth of the physical communication channel is effectively utilized.

Figure 6.9 illustrates the test results from the experimental examination of the H.264-10 High Quality compression method generated frame size dependency on the resolution of the Standard clip used. The experiment uses the basic Standard clip shot at a resolution of 2600 × 1950 dots and clip modification by resolution reduction to QVGA (320 × 240 dots) (Figure 6.10). The information on Figures 6.9 and 6.10 shows that with lower resolutions frame size increases relatively slower (Figure 6.11) and at megapixel resolutions increases from 10 to 70 KB.

Figures 6.10 and 6.12 provide experimental results when comparing two of the most widely used video compressing algorithms — H.264 and MJPEG (MPEG, especially, is widely used for real time camera picture visualization over a browser or WWW). Comparison of the two methods concerning the two basic network oriented parameters — frame size and bandwidth used is presented on Figures 6.13 and 6.14.

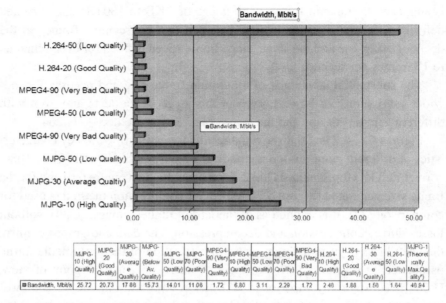

	MJPG-10 (High Quality)	MJPG-20 (Good Quality)	MJPG-30 (Average Quality)	MJPG-40 (Below Av. Quality)	MJPG-50 (Low Quality)	MJPG-70 (Poor Quality)	MPEG4-90 (Very Bad Quality)	MPEG4-10 (High Quality)	MPEG4-50 (Low Quality)	MPEG4-70 (Poor Quality)	MPEG4-90 (Very Bad Quality)	H.264-10 (High Quality)	H.264-20 (Good Quality)	H.264-30 (Average Quality)	H.264-50 (Low Quality)	MJPG-1 (Theoretically Max.Quality*)
Bandwidth, Mbit/s	25.72	20.73	17.86	15.73	14.01	11.06	1.72	6.80	3.11	2.29	1.72	2.46	1.88	1.58	1.64	46.94

Fig. 6.6 Comparison of test results for consumed Bandwidth by the different methods of video compression for the Reference input video stream with the following parameters: Resolution — Full HD; Number of frames per second — 10; complexity of the images — 50%; movement or dynamics of the live scene — 50%.

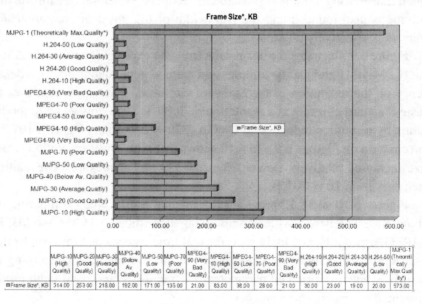

	MJPG-10 (High Quality)	MJPG-20 (Good Quality)	MJPG-30 (Average Quality)	MJPG-40 (Below Av Quality)	MJPG-50 (Low Quality)	MJPG-70 (Poor Quality)	MPEG4-90 (Very Bad Quality)	MPEG4-10 (High Quality)	MPEG4-50 (Low Quality)	MPEG4-70 (Poor Quality)	MPEG4-90 (Very Bad Quality)	H.264-10 (High Quality)	H.264-20 (Good Quality)	H.264-30 (Average Quality)	H.264-50 (Low Quality)	MJPG-1 (Theoretically Max Quality*)
Frame Size*, KB	314.00	253.00	218.00	192.00	171.00	135.00	21.00	83.00	38.00	28.00	21.00	30.00	23.00	19.00	20.00	573.00

Fig. 6.7 Comparison of test results for the Frame size generated by the application of different methods of video compression for the Reference input video stream with the following parameters: Resolution — Full HD; Number of frames per second — 10; complexity of the images — 50%; movement or dynamics of the live scene — 50 %.

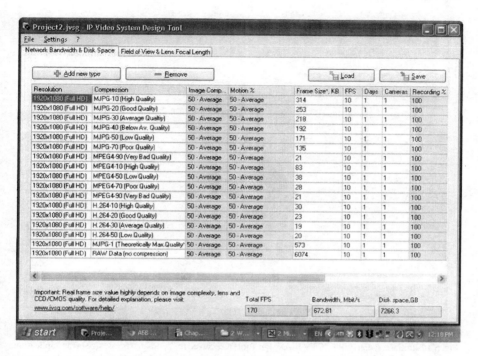

Fig. 6.8 Test results from CCTV IP-Design Tool™, obtained with the different methods of video compression for the Reference input video stream with the following parameters: Resolution — Full HD; Number of frames per second — 10; complexity of the images — 50%; movement or dynamics of the live scene — 50%.

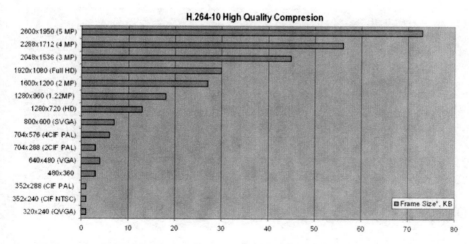

Fig. 6.9 Prediction results from CCTV IP-Design Tool™, received for the generated Frame size, in the application of compression method H.264-10 High Quality at various resolutions of the Reference input video stream: Resolution — according to the figure; Number of frames per second — 10; complexity of the images — 50%; movement or dynamics of the live scene — 50%.

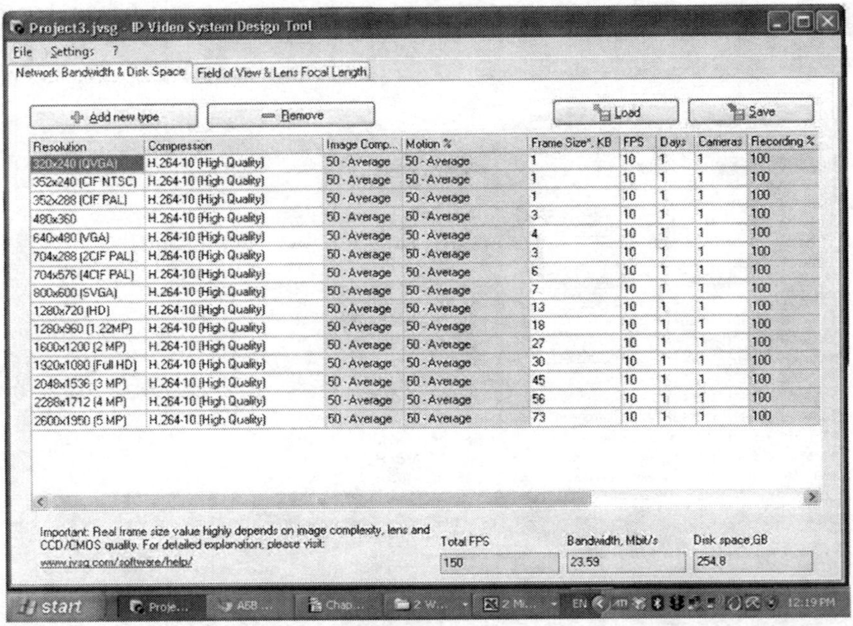

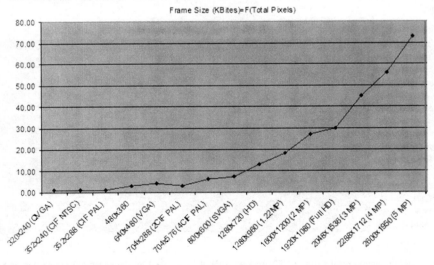

Fig. 6.10 Initial test data from CCTV IP-Design Tool™, received the generated frame size in the application of compression method H.264-10 High Quality at various resolutions of the Reference input video stream: Resolution — according to the figure; Number of frames per second — 10; complexity of the images — 50%; movement or dynamics of the live scene — 50%.

Frame Size (KBites)=F(Total Pixels)

Fig. 6.11 Relationship between resolution and frame rate obtained for the generated frame size in the application of compression method H.264-10 High Quality at various resolutions of the Reference input video stream: Resolution — according to the figure; Number of frames per second — 10; complexity of the images — 50%; movement or dynamics of the live scene — 50%.

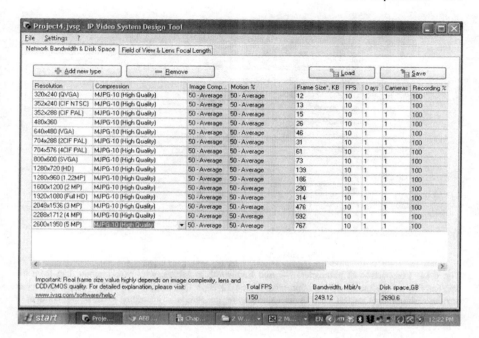

Fig. 6.12 Initial test data from CCTV IP-Design Tool™, received the generated frame size when applying the method of compression MJPEG-10 High Quality at various resolutions of the Reference input video stream: Resolution — according to the figure; Number of frames per second — 10; complexity of the images — 50%; movement or dynamics of the live scene — 50%.

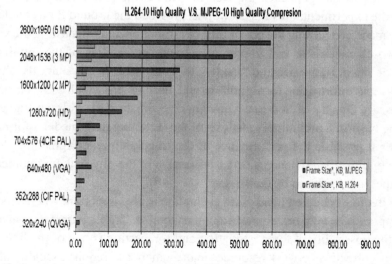

Fig. 6.13 Comparison of parameters — frame size in KB, in experimental data from the applicability of the methods of compression H.264-10 High Quality and MJPEG-10 High Quality at various resolutions of the Reference input video stream: Resolution — according to the figure; Number of frames per second — 10; complexity of the images — 50%; movement or dynamics of the live scene — 50%.

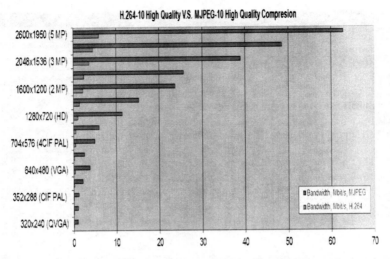

Fig. 6.14 Comparison of parameters — used Bandwidth in Mbit/s, with experimental data from the applicability of the methods of compression H.264-10 High Quality and MJPEG-10 High Quality at various resolutions of the Reference input video stream: Resolution — according to the figure; Number of frames per second — 10; complexity of the images — 50%; movement or dynamics of the live scene — 50%.

6.11 Conclusions

This chapter has presented an overview and comparative analysis, together with the experimental results, as well as provided the ground for a conclusion to be made concerning the possibility of estimation and forecasting the values of two characteristic and important network parameters of the multimedia type of video: frame size and 'bandwidth' consumed. In this sence the chapter gives input information for algorithms and mechanisms which could ensure the necessary Quality of Service. Knowing or estimating the average frame size on the basis of a preliminary analysis of the incoming multimedia information makes it possible the performance of the fragmenting/defragmenting logics to be optimized and hence a better utilization of the physical communication channel bandwidth to be realized.

Effective occupied bandwidth evaluation is also important when choosing an approach to network connection management. Traffic parameters analysis and the possibility of their estimation and forecasting is one of the first steps towards effective network resources management and required Quality of Service provision. The accumulation of statistical information and building up of a data base with frame size values and estimations of required bandwidth are a condition for the implementation of adaptive management procedures of the network connections and provision of the required network Quality of Service when transmitting multimedia information streams.

References

[1] N. Chapman and J. Chapman, Digital Multimedia, John Wiley and Sons Ltd, 2000.

[2] H. Schulzrinne, S. Casner, R. Frederick, and V. Jacobson, "RTP: A Transport Protocol for RealTime Applications", RFC1889, January 1996.

[3] B. Furht, *Multimedia Tools and Applications*, Kluwer Academic Publishers, 1996.

[4] J. Leigh, O. Yu, D. Schonfeld, and R. Ansari, et. al., "Adaptive networking for tele-immersion", *Proc. Immersive Projection Technology/Eurographics Virtual Environments Workshop (IPT/EGVE)*, Stuttgart, Germany, 2001.

[5] S. Kent and R. Atkinson, "Security Architecture for the Internet Protocol", RFC2401, November 1998.

[6] V. K. Garg and O. T. W. Yu, "Integrated QoS support in 3G UMTS networks", *IEEE Wireless Communications and Networking Conference*, vol. 3, pp. 1187–1192.

[7] H. Schulzrinne and J. Rosenberg, "The Session Initiation Protocol: Internet-Centric Signaling", *IEEE Communications Magazine*, October 2000.

[8] D. Bertsekas and R. Gallager, *Data Networks*, Prentice Hall, 1987.

[9] A. Tanenbaum, *Computer Networks*, 3e, Prentice Hall, 1996.

[10] A. Leon-Garcia and I. Widjaja, *Communication Networks: Fundamental Concepts and Key Architectures*, McGraw-Hill, 2000.

[11] R. Arora and R. Jain, "Voice over IP: Protocols and Standards", Class report avail. on-line at ftp://ftp.netlab.ohio-state.edu/pub/jain/courses/cis788-99/voip_protocols/index.html

[12] H. Liu and P. Mouchtaris, "Voice over IP Signaling: H.323 and Beyond", *IEEE Communications Magazine*, October 2000.

[13] B. Cavusoglu, D. Schonfeld, and R. Ansari, "Real-time adaptive forward error correction for MPEG-2 video communications over RTP networks", *Proceedings of the IEEE International Conference on Multimedia and Expo*, Baltimore, Maryland, 2003.

[14] R. Oppliger, "Security at the Internet Layer", *Computer*, vol. 31(9), September 1998.

[15] D. Schonfeld, "Image and Video Communication Networks", (Invited Chapter). Handbook of Image and Video Processing. A. Bovik (ed.), Academic Press: San Diego, California, Chapter 9.3, pp. 717–732, 2000.

[16] J. Kurose and K. Ross, *Computer Networking: A top-down approach featuring the Internet*, Addision Wesley, 2001.

[17] G. A. Thom, "H.323: the multimedia communications standard for local area networks", *IEEE Communications Magazine*, vol. 34(12), December 1996, pp. 52–56.

[18] M. Banikazemi and R. Jain, "IP Multicasting: Concepts, Algorithms, and Protocols", Class report avail. on-line at http://ftp.netlab.ohio-state.edu/pub/jain/courses/cis788-97/ip_multicast/index.htm

[19] L. H. Sahasrabuddhe and B. M. Mukherjee, "Multicast Routing Algorithms and Protocols: A Tutorial", *IEEE Network*, January/February 2000.

[20] A. Striegel and G. Manimaran, "A Survey of QoS Multicasting Issues", *IEEE Communications Magazine*, June 2002.

[21] A. B. Johnston, Understanding the Session Initiation Protocol, Artech House, 2000. Multimedia Networks and Communication.

[22] E. Rosen, A. Viswanathan, and R. Callon, "Multiprotocol Label Switching Architecture", RFC3031, January 2001.

[23] N. Tang, S. Tsui, and L. Wang, "A Survey of Admission Control Algorithms", on-line report available at http://www.cs.ucla.edu/~tang/

[24] P. P. White, "RSVP and integrated services in the Internet: a tutorial", *IEEE Communications Magazine*, vol. 35(5), May 1997, pp. 100–106.

[25] S. Blake, D. Black, M. Carlson, and E. Davies, et al., "An Architecture for Differentiated Services", RFC2475, December 1998.

[26] B. P. Crow, I. Widjaja, J. G. Kim, and P. T. Sakai, "IEEE 802.11: Wireless Local Area Networks", *IEEE Communications Magazine*, September 1997.

[27] F. Kuipers, P. V. Mieghem, T. Korkmaz, and M. Krunz, "An Overview of Constraint-Based Path Selection Algorithms for QoS Routing", *IEEE Communications Magazine*, December 2002.

[28] E. Mulabegovic, D. Schonfeld, and R. Ansari, "Lightweight Streaming Protocol (LSP)", *ACM Multimedia Conference*, Juan Les Pins, France, 2002.

[29] R. Braden, L. Zhang, S. Berson, S. Herzog, and S. Jamin, "Resource ReSerVation Protocol (RSVP): Version 1 Functional Specification", RFC2205, September 1997.

[30] J. Watkinson, The MPEG Handbook: MPEG-I, MPEG-II, MPEG-IV, Focal Press, 2001.

[31] D. Salomon, Data Compression: The Complete Reference, Springer, 1998. Multimedia Networks and Communication.

[32] V. K. Garg, IS-95 CDMA and cdma2000, Prentice Hall, 1999.

[33] S. Dixit, Y. Guo, and Z. Antoniou, "Resource Management and Quality of Service in Third Generation Wireless Networks", *IEEE Communications Magazine*, February 2001.

[34] W. Kinsner, "Compression and its metrics for multimedia", *Proceedings of First IEEE International Conference on Cognitive Informatics*, 2002, pp. 107–121.

Biography

Professor Vladimir Poulkov PhD, has received his MSc and PhD degrees at the Technical University of Sofia. He has more than 30 years of teaching and research experience in the field of telecommunications. The major fields of scientific interest are in the field of information transmission theory, modulation and coding. His has expertize in the field of interference suppression, power control and resourse management for next generation telecommunications networks. Currently he is Dean of the Faculty of Telecommunications at the Technical University of Sofia. Member of IEEE.

Associated Professor Oleg Asenov PhD, has received his MSc degree at the Technical University of Gabrovo and PhD degree at the Technical University of Sofia. He has more than 20 years of teaching and research experience in the field of networking and telecommunications. The major fields of scientific interest are computer networks modeling, simulation and design based on graph theory and applied heuristics algorithms. Currently he is the full time Associated Professor at the St.Cyril and St.Methodius University of Veliko Tyrnovo, Member of IEEE, Associated Member of the Auditors Body of HIPAA — CHRA Associated Risk Assessment Auditor.

7

Potential Applications and Research Opportunities in the CONASENSE Initiative

Mehmet Şafak

Telecommunications Group, Hacettepe University, Turkey

7.1 Introduction

Recent advances in digital communications and high-speed digital signal processing led to innovative technologies, techniques, systems and services in the areas of communications, navigation and sensing. These changes, supported by the integration of transmission of voice, data and video by using Internet Protocol (IP) and the accompanying increase in the demand for these services, greatly improved the versatality, availability and ubiquitous use of services in this domain.

Among these, the demand is constantly increasing for services related to positioning, tracking and navigation of some users/platforms. For example, we currently use available services for determining and tracking the position of a user in mountanous/forest areas or in seas, for finding the address of a colleague that we want to visit as we drive in a large city or learning the status of a parcel in a postal service.

Similarly, we observe an unprecented development in sensing technology, sensors and sensor networks. A variety of sensor types are now available on

COmmunications- NAvigation-SENsing-SErvices (CONASENSE), 143–164.

the market in many domains, from tasting the quality of wine/tea/coffee to determining the temperature, the humidity and the mineral and water content of the ground for agricultural purposes, sensing/monitoring the physiological conditions of drivers/patients etc. Sensors operate at various frequency bands and locations, e.g., indoor/outdoor, airborne, space-borne, terrestrial, underwater and underground.

Traditional approaches for the provision of these services, e.g., the allocation of different frequency bands and waveforms and hence different receiver platforms for these services may not be optimal for integration. For example, the navigation signals generally contain information only about the platform identity, the platform location and the transmission time. The bandwidth allocated to only navigation services should be sufficiently large so as to allow accurate position determination. In view of this, and the fact that modern telecommunication systems support very high data rates, integration of navigation and telecommunication services may be feasible.

The CONASENSE initiative is about the cross-fertilization of digital COmmunications, NAvigation, SENsing and ensuing SErvices. Telecommunications provide the infrastructure for the provision and the ubiquotous use of services related to navigation and sensing (see Figure 7.1). The technology is available for the integration of services related to communications, sensing and navigation (CONASENSE) under realistic integration scenarios.

Covering a large domain of research, the CONASENSE has a high potential for a variety of applications and service provision to a large spectrum of users. Consequently, the CONASENSE-related studies may be research-oriented, e.g., related to system architecture, performance evaluation, protocol design,

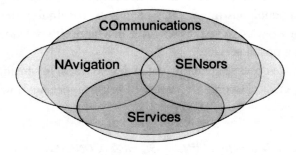

Fig. 7.1 The CONASENSE framework.

physical layer techniques etc.; application-oriented, e.g., proof-of-concept studies, system or prototype development; and/or service-oriented, including approaches for the provision of a multitude of services. These services may be related to, for example, health, monitoring and protection of the environment, traffic-control and security.

This survey does not claim to be exhaustive and aims only to provide a flavour about the requirements, research and application areas related to the CONASENSE-related services. The requirements for the present and future CONASENSE services will be briefly outlined in Section 2. Section 3 will review the potential research areas in positioning and sensing, where telecommunications is always omnipresent. Finally, Section 4 will outline some selected CONASENSE applications, such as e-Health, security, traffic control, environment monitoring and protection.

7.2 Requirements

Below is presented a brief description of the requirements about the CONASENSE-related services from different perspectives.

7.2.1 Requirements for Terminals/Platform

The users, transceivers, and the platforms to be navigated, and the targets/parameters to be sensed in-situ or remotely may be listed as space-borne (satellites, unattended aerial vehicles (UAVs), high-altitude platforms), sea-borne, ground-based, underground (earthquake, mines, tunnels etc.), underwater (communications, vehicle navigation, etc.), isolated places (mountains, forests, seas, oceans, deserts), indoor/outdoor etc. For example, space-borne telescopes are already being used successfully in radio astronomy. Near-earth orbiting satellites are used in many areas such as remote-sensing, harvest prediction, surveillance and environmental protection. Sensing and telecommunications may also be used for safety purposes, e.g., monitoring disaster and undeground mines and tunnels. Similarly, underwater communications, sensing and navigation may have numerous application areas, including navigation of underwater platforms, remote sensing wild life in the seas/oceans/deserts, fishing industry, mineral exploration in the oceans and monitoring and protection of the environment.

7.2.2 User Requirements

The requirements for CONASENSE-related services are mainly related to the user mobility and the environment. The users would prefer to have low-cost, light-weight, user-friendly, and low power/energy consuming receiving terminals, operating at all/most frequency bands allocated to the CONASENSE-related services with optimized coverage. Availability of these services may impose stringent requirements, e.g., the availability of navigation services might be of vital importance in certain scenarios. The privacy and survivability may be required for many services. The cognitivity, self-organization capability and adaptivity of the terminals against different environmental and operational conditions are strongly desired.

7.2.3 Requirements for Receiver and System Design

The frequency bands allocated to CONASENSE services differ depending on the particular application. Consequently, the propagation channels behave differently and the requirements for transceiver design are not the same. Depending on the applications and the related services, the operating frequency bands may cover radio frequency (RF), optical, infrared (IR), acoustics etc. A common (integrated) and interoperable receiver for CONASENSE services is required to operate at different frequency bands, transmit powers, receiver sensitivities, antenna structures, single- and multi-carrier transmissions techniques and modulation schemes [1–7]. Improved positioning accuracy by integration of the navigation data collected from different sources such as global positioning system (GPS), Galileo, GLONASS, Wi-fi, gyroscopes, accelerometers etc. is very much desirable.

Software-defined radio (SDR) may be considered as a strong candidate for an integrated receiver. Present technologies allow the design of agile front-ends with frequency synthesizers of fast hopping and settling times and with low phase noise. Followed by fast analog-to-digital and digital-to-analog convertors (ADC/DACs), the present signal processing technology facilitates the design of common low-noise and sensitive SDR receivers, by using a DSP chip, ASIC or FPGA, operating at numerous frequency bands and interoperable with different systems [8, 9].

In view of the above, the architectural system design which would enable us to provide integrated services in the domain of CONASENSE is a challenging task. One of the first issues to be resolved in this context concerns the interoperability with LTE/4G, professional mobile radio (PMR) and/or terrestrial and satellite-based navigation systems [7]. In view of the anticipated diverse applications, direct communications and multi-hop relaying between the equipments of different users without the need for base stations may also be required. Other issues to be concerned include broadcasting, multicasting, security, routing, radio resource management, efficient power/energy consumption, network establishment etc.

7.3 Potential Research Areas

Communications provide infrastructure for navigation- and sensing-related applications and availablility of these services to the users. This section will provide a brief summary of potential research areas on positioning, navigation and sensing and their relations with telecommunications in the near- to mid-term. Long-term predictions are believed to be more difficult since they are closely related to very fast advances that we observe in the above-mentioned domains.

7.3.1 Positioning

Positioning systems emit signals to help user terminals to determine time, their coordinates (latitude longitude and altitude from sea level), as well as their speed, accelaration and direction of movement. Positioning information is used in telecommunications, e.g., in the formation and self-organization of ad hoc networks [10], and in navigation and sensing [11, 12]. Accurate, reliable and real-time positioning is a serious issue in the operation of location-aware services and depends on the availability of initial accurate position information and motion measurement data. *Position estimation* refers to the mobile positioning problem when both the initial location and motion measurement data are not available. In that case, the geolocation of a mobile terminal is estimated using location-dependent signal parameters, usually using the signals exchanged between the mobile and a number of reference terminals in challenging propagation conditions such as multipath fading and shadowing. When only motion measurements are available, the problem is known as *global localization* [11]. If both the initial location and motion measurement data are available, the positioning problem is referred to as *position tracking*. Hence, position estimation is usually a prerequisite for position tracking.

Positioning systems are usually categorized as network-based or mobile-based depending on the location where position calculations are performed. Calculations for position estimation may be mobile-based, when the necessary information is extracted from the received signals, or network-based if information collected through reference terminals is processed at a central unit.

Positioning systems may be either terrestrial-based and used for both outdoor and indoor environments, or satellite-based, which offer global coverage but generally serve to only outdoor users. In terrestrial-based systems either geometric or mapping techniques are usually applied. In geometric positioning, the signal parameters, such as time-of-arrival, angle-of-arrival, time difference-of-arrival, and received signal strength, are directly used for position estimation. Mapping-based mobile positioning employs geometric techniques and a database of location-dependent parameters, which are obtained by theoretical channel models and/or measurements. Mapping techniques provide improved positioning accuracy, because no LOS conditions are needed and may operate with even one reference source. All the above methods differ in terms of accuracy, coverage, cost, mobile terminal power consumption and their impacts on the wireless system [12].

Commonly used terrestrial navigation systems are LORAN-C, which is widely used in the Western world, and its Russian counterpart CHAYKA. These systems enable ships and aircraft to determine time, their position and speed from low-frequency radio signals emitted by fixed land-based radio beacons. For example, by using accurate cesium clocks, LORAN-C disseminate 5 MHz and pulse per second (pps) signals from high power transmitters to allow navigation receivers to synchronize with typically ±100 ns accuracy.

The so-called global navigation satellite systems (GNSSs) provide geospatial positioning with global coverage. The CDMA signals transmitted by these systems enable receivers to determine time and their position (altitude, longitude and latitude) with accuracies in the order of 1 meter. GNSS systems use typically 24–35 medium earth orbit satellites with orbital heights of approximately 20000 km and periods typically 12 hours. Most commonly used GNSSs are GPS and its Russian counterpart MOLNYA. The GPS system has two variants: stand-alone GPS or assisted-GPS (A-GPS). The first is mobile-based, while A-GPS needs extra signals from reference GPS receivers and provide positioning accuracy on the order of centimeters. The European GNSS, Galileo, is expected to be fully operational by 2020 and be compatible with the GPS system. Similarly, the Chinese system COMPASS is expected to be fully operational by 2020.

GNSSs have global coverage, are mainly employed for outdoor applications, and yield higher position accuracy. However, the benefits of satellite-based positioning could be limited when satellite signals are shadowed and/or the receiving terminals cannot see at least four GPS satellites simultaneously. In such cases, one may resort to other positioning methods in order to backup the failed or degraded satellite signals; this requires a receiver operating with all these systems.

GPS may not be suitable for indoor use because of the unavailability of line-of-sight transmission between indoor GPS receivers and satellites. Compared with outdoor positioning, smaller distances are involved in indoor environments with highly variable and dynamic changes to deal with. Enabling technologies for indoor positioning systems include radio-frequency identification (RFID), wireless local area networks (WLANs), Bluetooth, sensor networks, ultra-wideband (UWB), IR, ultrasound, magnetic signals, vision analysis and audible sound [13, 14]. Indoor positioning systems may be successfully used in inventory tracking, home automation and health monitoring applications, guided tours in museums etc. [15, 16].

The micromechanical systems (MEMS) technology enables the development of giroscopes and accelerometers at smaller sizes to be incorporated in mobile terminals. Optimum fusion of the signals obtained by various sensors for improved positioning accuracy is still an active research area. The positioning accuracy for indoor and outdoor environments is expected to be improved drastically and pave the way for many innovative CONASENSE applications even in the near future.

The waveforms, operational frequencies and the capabilities of these systems are not the same. There are receivers available in the market which can process these signals jointly for time and position determination. However, there are still much to do for the integrated design of these systems for improved time and position estimation. For example, efficient infrastructure design of a hybrid terrestrial-satellite positioning system and its integration with communication systems is still a challenging problem. Design and production of multiband receiver antennas are required to operate in the frequency bands allocated to navigation and communication systems, and of transmit antennas that produce signals with isotropic power spectral density within global coverage for navigation receivers [17].

Channel modeling for terrestrial and satellite navigation systems is required for all operating frequencies and transmission bandwidths, types and locations of the navigation platforms and receiving terminals. The limitations imposed on time and position accuracy by multipath fading, shadowing, diffraction, ionospheric effects, outdoor to indoor coupling and other propagation effects requires careful consideration [18]. The need for advanced positioning techniques is believed to increase for improved performance of CONASENSE-related services. The use of multiple-input multiple-output (MIMO) techniques is expected to improve not only the performace of the communication systems but also help for accurate position and time estimation.

There is currently a strong interest in telecommunications, sensing and navigation communities to bio-inspired algorithms [19] for improving the performance of CONASENSE-related services. Evolution-perfected bio-algorithms for colony life, navigation and migration of fishes, bees [20], ants, birds and herds inspire the scientists to exploit bio-inspired algorithms more agressively.

7.3.2 Sensing

Recent advances in digital technology enabled the development and production of high resolution, low-power consumption, environment-friendly, long-life, low-cost and small-size sensors [21, 22]. Consequently, we observe in our everyday life various sensor types, including RFID-, MEMS-, biometric-, acoustic-, video-sensors and so on. Sensors are used in a very large spectrum of applications, including health monitoring [23, 24], underwater acoustic networks [25], smart grid applications [26], agriculture [27], and automative industry [28].

Sensors may be used for sensing locally or remotely; the information collected by the sensed signals may be processed in situ, in a distributed fashion, or at a fusion center [29, 30]. Multiple sensors may be employed for cooperative sensing when the data collected by a single sensor does not meet the requirements. Cooperative sensing by mobile sensors in wireless channels under fading and correlated shadowing is still among the topics of intense research efforts. Data processing architecture and techniques, e.g., sensor fusion, data fusion and/or information fusion, managing the information collected by sensing and making an optimum multi-criteria decision concerning the sensed data deserve careful consideration.

7.3.2.1 Electronic Nose, Eye and Tongue

Recent research efforts on electronic nose, electronic eye and electronic tongue is leading to versatile and innovative applications. Smelling is based on the interaction of odour molecules with an array of gas sensors, leading to an electrical response carrying information about odour, context, freshness, quality etc. [31]. Among the application areas, one can cite aroma, colour and taste classification of tea, coffee, beer, wine, spices, monitoring the ripeness and freshness of fruits and vegetables and in other areas of food industry [32, 33]. Location of odour sources and online monitoring of livestock farm odours may be used in farms [34]. Sensing and monitoring volatile organic compounds, atmospheric pollution, hazardous gases, chemicals and explosives may be cited among the applications for security and environmental protection. In health-related applications, one may cite the diagnosis of lung cancer at early stage [35], identification of urinary tract infection [36] and helicobacter as well as detection, discrimination and monitoring of drug, drug users and smokers. Sensor systems are also used for building artificial nose, tongue and eye for robots and other applications as cited in the literature [37–39].

Cognitive and biologically inpired algorithms, machine learning, fuzzy and neural techniques, and graph theory are currently among the areas of interest for vision research and electronic eye. Decision algorithms used in sensor networks include Bayesian classifier approach, rough set-based approach, fuzzy approach, singular value decomposition and pattern classification [40, 41].

7.3.2.2 Smart Power Grid

Smart power grid modernizes the current electricity delivery system by integrating information and communications technologies (ICT) into generation, delivery, control and consumption of electrical energy for enhanced robustness against failures, efficiency, flexibility, adaptivity, reliability and cost-effectiveness. In that sense, smart grid embodies a fusion of different technologies where electrical power engineering meets sensing, ICT, positioning, control etc. [42]. The communication protocols to be deployed for this purpose should modernize and optimize the operation of the power grid and provide an accessible and secure connection to all network users, especially for efficient use of distributed energy sources [43]. This is achieved through automated control, high-power converters, energy storage installations, modern communications infrastructure, sensing and metering technologies, and modern energy management techniques based on the optimization of demand, energy and network availability and more. Through communications between interconnected sensor nodes, the smart grid controls equipment and energy distribution, isolates and restores power outages, facilitates the integration of renewable energy sources into the grid and allows users to optimize their energy consumption. Due to its high potential for economic, environmental, and societal benefits, the development and implementation of the smart power grid is considered to have global priority [44–47].

Design of end-to-end QoS resource control architectures [48], efficient schemes for admission control, monitoring/control of the smart grid and the fluctuations of the power load is believed to be of prime importance for smart power grid networks [49]. Reliability and security of integrated communication network of the Smart Grid should be guaranteed in very adverse power line channels, which suffer high attenuation, multipath, impulse noise, harmonics and distortion [50]. In addition to more reliable power line channel models [48], robust modulation, coding, encryption [42, 52] and transmission techniques and computational models are needed for optimized physical layer performance.

New business models and policy designs are required in view of the competitive privatized power market.

The available technologies related to entertainment, home automation and e-Health present different operational characteristics and the related systems are in general not compatible or interoperable. Entertainment requires high-bandwidth and real-time streaming; home automation requires low-power and small sensors; while e-Health domain requires high-security and ubiquity. Recent advances in consumer networking are promising for the integration of these services with relevant architectural approaches [53].

7.4 Potential Applications

7.4.1 Health

In view of rising costs in healthcare, increase in the percentage of ageing population and the fact that many patients needing health-monitoring do not necessarily require hospitalisation, new technologies related to CONASENSE may play a major role in the provision of health services, especially for disabled, elderly and chronically ill patients at lower costs [54–58].

The progress in ICT, biotechnologies and nanotechnologies accelerate innovations in the field, and lead to miniaturization and large-scale production of efficient and affordable products. Standardization will ensure interoperability between devices and information systems, and open up the way for large-scale and cost-effective deployment of e-health systems [59].

Thanks to recent advances in sensor technology and networks, the human health can be monitored by collecting data on specific physiological indicators (e.g., blood glucose level, blood pressure, electrocardiogram and electroencephalogram, portable magnetic resonance images, implantable hearing aid etc.), via in-, on-, and/or out-body sensors. Such systems typically perform sensing, data collection with user profile information, including data aggregation, data visualization, and analysis/alerting functions for the health professionals. The data, which is usually collected at a hub, is periodically transmitted to a server through a gateway using IP. The database in the server may be used for preventative health care, physiological/functional monitoring, chronic disease management, and assessment of the quality of life, e.g., fitness, diet or nutrition monitoring applications. In view of the multiplicity and mobility of sensors and users, association of the health monitoring data with patients requires serious consideration [60].

In e-health systems, the patients or sensing systems may update their data in real time through Internet. Hence, the record of a patient becomes available to authorized professionals anytime anywhere, for real-time monitoring and intervention in emergency. Such systems also provide new forms of interaction and coordination between health professionals and lead to novel scientific approaches for medical applications.

E-health systems provide mobility to the patients via mobile health-monitoring devices [61]. Patient mobility also calls for wearable, outdoor and home-based applications. Physical and health conditions of patients can be monitored in real-time using sensors observing the environment and those that measure physiological parameters of the patient at home and hospital environments. Similar approaches may be followed for the safety of workers [62]. Indoor positioning and tracking systems may be used in hospitals, for example to track expensive equipments, and to guide patients and health professionals inside the hospitals for more efficient and timely services.

The health data can be stored on the sensor nodes and analyzed offline, while emergency situations and reports may be made available to health workers through a remote database. Hence, stored data from multiple patients may be utilized for geographic and demographic analyses [63]. Mobility solutions for wireless body area networks (BANs) for healthcare are already available, through communications over Low-power Personal Area Networks (6LoW-PANs) using Internet protocol Ipv6 [64]. These systems can provide preventive healthcare, enhanced patient–doctor interaction and information exchange. Continuous health-monitoring allows immediate intervention in case of an emergency. Positioning, tracking and monitoring patients with asthma, diabetics, heart disease, alzheimer, obesity [65] as well as visually impaired people, ambulance systems, small children, robotic wheelchairs, pregnancy, blood pressure, artificial arms/legs, drug addiction are believed to be important issues.

Modelling the physical channel in BANs is the topic of intense research efforts [66–68]. Optimal use of relay nodes, adaptive approaches for managing outages and retransmissions, cross-layer optimization to share information between physical and media-access control (MAC) layers will definitely improve the overall system performance. The MAC-layer operation in BANs for e-health is addressed in the IEEE 802.15.6 draft standard for BAN [69]. For example, intra-body communication for continuous-monitoring of patients with artificial heart embodies serious issues, e.g., the operation of an antenna and the propagation of electromagnetic waves in human body, which is a non-homogeneous lossy medium. Similarly, reliable signal transmission between sensors on the body under shadowing is still among the areas of interest. Positioning with high precision in the human body, which is strongly desired for surgical operations, is believed to be possible in the near/mid-term.

The physiological data collected in e-health systems is bidirectional and distributed through Internet and/or in heterogeneous networks to all interested parties. Nevertheless, the problems related to architecture and protocol design, accuracy, reliability, data security, protection, privacy of diagnoses, range, operation time, and interoperability between medical devices are still of research interests [70–72]. These studies should be supported by databases, intelligent decision support algorithms and programming languages. Recent efforts on medical device interoperability resulted in a standard ISO/IEEE 11073 PHD [73] for communications between health devices such as USB, Bluetooth, and ZigBee.

CONASENSE paves the way for wireless e-health applications including healthcare telemetry and telemedicine [74]. Even the patients at some remote locations and/or unable to reach a nearby health center may be monitored and managed; remote diagnosis and emergency intervention can be accomplished by tele-medecine. Improved and low-cost healthcare services may be provided to poor and geographically remote patients by exploiting new technologies. In that context, such services may provide a low-cost healthcare solution in less developed geographic regions in the world.

7.4.2 Security

PMR systems, e.g., APCO and TETRA, are employed by police, fire departments, ambulance systems etc. Compared with ubiquitous commercial mobile radio systems, these systems have some additional requirements for survivability in disaster scenarios, operation in relay mode without needing base stations, larger coverage areas (higher transmit powers) etc. Interoperability between mobile radio and PMR systems is strongly desired. 4G systems may provide services for virtual private networks to the PMR users, since the design, operation and maintenance of two separate systems are costly. With present technology, features like relaying and survivability may be provided through diversity and coordination between base stations. Incorporation of sensing capability and accurate time- and positioning estimation in mobile radio systems may put them in a very strong position for monitoring, and management of disaster, emergency, mine-accidents, earthquake, fire-fighting, police patrolling, and intruder/fraud detection. Rapid and accurate position estimation/navigation is very often needed in military applications [11].

With accurate position estimation, a variety of applications and services, such as location sensitive billing, and improved traffic management for cellular networks can become feasible. Positioning of a mobile terminal is considered to be critical for position-aware services such as such as E-911 in USA and E-112 in European Union (EU) for emergency calls [75, 76]. Noting that mobile-originated emergency calls are continually increasing and about 50% of all emergency calls in the EU are originated by mobile users, location estimation of a mobile user making an emergency call is strongly desired.

7.4.3 Traffic Control

CONASENSE may also have a significant contribution in the domain of traffic control. This covers a large area of research and applications. For example, monitoring and management of highway/tunnel/bridge traffic which may need to be diverted, under congestion, to alternative itineraries may save valuable money and time and reduce pollution. Intelligent transport systems (ITS) will significantly alleviate the urban traffic via inter-vehicular communications, communications between the terminals along the roads, and broadcasting to vehicles the last-minute traffic information. Controlling the distance between vehicles on the road by radars on the vehicles for traffic under rain/snow/fog is proved to be very effective [77–82]. Similarly monitoring and navigation of the traffic in railways, harbors/ports/seas [83], e.g., yachts, ships etc., air traffic (air trafic control and taxi) and UAV's are among the application areas of the CONASENSE. Monitoring and controlling the border traffic between countries may also be used to prevent illegal border crossings. Traffic control in shopping malls, banks, concert halls etc. may be desirable for statistical purposes as well as for security reasons. Concepts for traffic control, e.g., monitoring the migration paths, times and the density of wild animals, could be valuable for the protection of wild life. Similar arguments may be repeated for the monitoring of the farm animals.

7.4.4 Environment Monitoring and Protection

Rapid growth of the world population, high cost of transforming already established and highly polluting manufacturing plants to become more environment-friendly constitutes serious threats to our planet. Fortunately, recent advances in CONASENSE-related technologies enable us to follow more environment-friendly approaches at lower costs. Higher resolutions in positioning and remote sensing is promising for monitoring and assessment of earth resources, agricultural harvest, forests, seas, wild life, water resources, weather/climate, ozon layer, electromagnetic and chemical pollution.

7.5 Conclusions

On the one hand, recent advances on sensor technologies contribute to the design and developments of more sensitive smaller, low-power and more-capable sensors. On the other hand, communication and digital signal processing techniques paved the way for drastic improvements in the performance of sensor networks, including the sensor fusion techniques. In addition, in the near/mid term, it is highly likely to foresee drastically improved positioning accuracies than at present. In view of the above, one can confidently anticipate very innovative, diversified, cheaper and reliable CONASENSE-related services due to ever increasing integration of communication, navigation and sensors.

References

Software-defined radio

[1] Bagheri, R., et al., "Software-Defined Radio Receiver: Dream to Reality," *IEEE Communications Magazine*, Vol. 44, No. 8, August 2006, pp. 111–118.
[2] Valls, J., T. Sansaloni, A. Pérez-Pascual, V. Torres, and V. Almenar, "The Use of CORDIC in Software Defined Radios: A Tutorial," *IEEE Communications Magazine*, Vol 44, No. 9, September 2006, pp. 46–50.
[3] Minde, G. J., et al., "An Agile Radio for Wireless Innovation," *IEEE Communications Magazine*, Vol. 45, No. 5, May 2007, pp. 113–121.
[4] Björkqvist, J. and S. Virtanen, "Convergence of Hardware and Software in Platforms for Radio Technologies," *IEEE Communications Magazine*, Vol. 44, No. 11, November 2006, pp. 52–57.
[5] Zhao, Y. et al., "A Software Defined Radio Receiver Architecture for UWB Communications and Positioning," *Canadian Conf. Electrical and Computer Engineering* (CCECE), 2006, pp. 255–258.
[6] Ihmig, M. and A. Herkersdorf, "Flexible Multi-Standard Multi-channel System Architecture for Software Defined Radio Receiver," *9th Int. Conf. on Intelligent Transport Systems Telecommunications, (ITST)*, 2009, pp. 598–603.
[7] Haskins, C. B. and W. P. Millard, "Multi-band Software Defined Radio for Spaceborne Communications, Navigation, Radio Science, and Sensors," *IEEE Aerospace Conf.*, 2010, pp. 1–9.
[8] Alluri, V. B., J. R. Heath, and M. Lhamon, "A New Multichannel, Coherent Amplitude Modulated, Time-Division Multiplexed, Software-Defined Radio Receiver Architecture, and Field-Programmable-Gate-Array Technology Implementation," *IEEE Trans. Signal Processing*, Vol. 58, No. 10, 2010, pp. 5369–5384.

[9]   Giannini V., et al., "A 2-mm 0.1–5 GHz Software-Defined Radio Receiver in 45-nm Digital CMOS," *IEEE Journal of Solid-State Circuits*, Vol. 44, No. 12, December 2009, pp. 3486–3498.

Positioning

[10] Mayorgaet, C. L. F., et al., "Cooperative Positioning Techniques for Mobile Localization in 4G Cellular Networks," *IEEE Int. Conference on Pervasive Services*, 2007, pp. 39–44.

[11] Khalaf-Allah, M., "Nonparametric Bayesian Filtering for Location Estimation, Position Tracking, and Global Localization of Mobile Terminals in Outdoor Wireless Environments," *EURASIP Journal on Advances in Signal Processing*, Vol. 2008, January 2008, pp. 1–14.

[12] Gezici, S., "A Survey on Wireless Position Estimation," *Wireless Personal Communications*, Vol. 44, 2008, pp. 263–282.

[13] Chiu, W.-Y., B.-S. Chen, and C.-Y. Yang, "Robust Relative Location Estimation in Wireless Sensor Networks with Inexact Position Problems," accepted for publication in *IEEE Trans. Mobile Computing*, 2011, DOI: 10.1109/TMC.2011.111.

[14] Chen, W., et al., "An Integrated GPS and Multi-Sensor Pedestrian Positioning System for 3D Urban Navigation," *Urban Remote Sensing Joint Event*, 2009, pp. 1–6.

[15] Zheng, V. W., J. Zhao, Y. Wang, and Q. Yang, "HIPS: A Calibration-less Hybrid Indoor Positioning System Using Heterogeneous Sensors," *IEEE Int. Conf. Pervasive Computing and Communications (PerCom)*, 2009, pp. 1–6.

[16] Gu, Y., A. Lo, and I. Niemegeers, "A Survey of Indoor Positioning Systems for Wireless Personal Networks," *IEEE Communications Surveys & Tutorials*, Vol. 11, No. 1, First Quarter 2009, pp. 13–32.

[17] Roederer, A. G., "Antennas for Space: Some Recent European Developments and Trends," *18th Int. Conf. Applied Electromagnetics and Communications* (ICECom), 2005, pp. 1–8.

[18] Martellucci, A., and R. P. Cerdeira, "Review of Tropospheric and Multipath Data and Models for Global Navigation Satellite Systems," *3rd European Conf. Antennas and Propagation* (EuCAP), 2009, pp. 3697–3702.

Sensors

[19] Stauffer, A., D. Mange, and J. Rossier, "Design of Self-organizing Bio-inspired Systems," *Second NASA/ESA Conference on Adaptive Hardware and Systems* (AHS), 2007, pp. 413–419.

[20] Bitam, S., M. Batouche, and E.-G. Talbi, "A Survey on Bee Colony algorithms," *2010 IEEE Int. Symposium on Parallel and Distributed Processing, Workshops and PhD Forum (IPDPSW)*, 2010, pp. 1–8.

[21] Fowler, K., "Sensor Survey Results: Part 1. The Current State of Sensors and sensor Networks," *IEEE Instrumentation and Measurement Magazine*, Vol. 12, No. 1, February 2009, pp. 39–44.

[22] Fowler, K., "Sensor Survey Results: Part 2. Sensors and Sensor Networks in Five Years," *IEEE Instrumentation and Measurement Magazine*, Vol. 12, No. 2, April 2009, pp. 40–44.

[23] Wang, C. H., Y. Liu, M. Desmulliez, and A. Richardson, "Integrated Sensors for Health Monitoring in Advanced Electronic Systems," *4th Int. Design and Test Workshop (IDT)*, 2009, pp. 1–6.

[24] Fernandez, J. M., J. C. Augusto, R. Seepold, and N. M. Madrid, "A Sensor Technology Survey for a Stress-Aware Trading Process," accepted for publication in *IEEE Trans. Systems, Man and Cybernetics- Part C: Applications and Reviews*, 2011.

[25] Garcia, J. E., "Positioning of Sensors in Underwater Acoustic Networks," *Proc. MTS/IEEE Oceans*, 2005, Vol. 3, pp. 2088–2092.

[26] Yang, Y., F. Lambert, and D. Divan, "A Survey on Technologies for Implementing Sensor Networks for Power Delivery Systems," *IEEE Power Engineering Society General Meeting*, 2007, pp. 1–8.

[27] Kalaivani, T., A. Allirani, and P. Priya, "A Survey on Zigbee Based Wireless Sensor Networks in Agriculture," *3rd Int. Conf. Trends in Information Sciences and Computing (TISC)*, 2011, pp. 85–89.

[28] Fleming, W. J., "Overview of Automotive Sensors," *IEEE Sensors Journal*, Vol. 1, No. 4, December 2001, pp. 296–308.

[29] Nicosevici, T., R. Garcia, M. Carreras, and M. Villanueva, "A Review of Sensor Fusion Techniques for Underwater Vehicle Navigation," *MTTS/IEEE Techno.- Oceans '04*, Vol. 3, 2004, pp. 1600–1605.

[30] Zhao, X., Q. Luo, and B. Han, "Survey on Robot Multi-sensor Information Fusion Technology," *7th World Congress on Intelligent Control and Automation (WCICA'2008)*, 2008, pp. 5019–5023.

Electronic nose, eye, and tongue

[31] Chang J. B. and V. Subramanian, "Electronic Noses Sniff Success," *IEEE Spectrum*, Vol. 45, No. 3, March 2008, pp. 51–56.

[32] Dutta, A., B. Tudu, R. Bandyopadhyay, and N. Bhattacharyya, "Black Tea Quality Evaluation Using Electronic Nose: An Artificial Bee Colony Approach," *IEEE Recent Advances in Intelligent Computational Systems (RAICS)*, 2011, Page(s): 143–146.

[33] Brezmes, J., et al., "Evaluation of an Electronic Nose to Assess Fruit Ripeness," *IEEE Sensors Journal*, Vol. 5, No. 1, 2005, pp. 97–108.

[34] Cai, J. and D. C. Levy, "Using Stationary Electronic Noses Network to Locate Dynamic Odour Source Position," *IEEE Int. Conf. Integration Technology (ICIT'07)*, 2007, pp. 793–798.

[35] Wang P., et al., "Development of Electronic Nose for Diagnosis of Lung Cancer at Early Stage," *Int. Conf. Information Technology and Applications in Biomedicine (ITAB 2008)*, 2008, pp. 588–591.

[36] Kodogiannis, V. S., J. N. Lygouras, A. Tarczynski, and H. S. Chowdrey, "Artificial Odor Discrimination System Using Electronic Nose and Neural Networks for the Identification of Urinary Tract Infection," *IEEE Trans. Information Technology in Biomedicine*, Vol. 12 , No. 6, 2008, pp. 707–713.

[37] Zhang, X., M. Zhang, J. Sun, and C. He, "Design of a Bionic Electronic Nose for Robot," *ISECS Int.l Colloq. on Computing, Communication, Control, and Management (CCCM '08)*, Vol. 2, 2008, pp. 18–23.

[38] Song, K., Q. Wang, H. Zhang, and Y. Cheng, "Design and Implementation a Real-time Electronic Nose System," *IEEE Instrumentation and Measurement Technology Conf. (I2MTC '09)*, 2009. pp. 589–592.

[39] Tang, K.-T., S.-W. Chiu, M.-F. Chang, C.-C. Hsieh, and J.-M. Shyu," A Low-Power Electronic Nose Signal-Processing Chip for a Portable Artificial Olfaction System", *IEEE Trans. Biomedical Circuits and Systems*, Vol. 5, No. 4, 2011, pp. 380–390.

[40] Bag, A. K., B. Tudu, J. Roy, N. Bhattacharyya, and R. Bandyopadhyay, "Optimization of Sensor Array in Electronic Nose: A Rough Set-Based Approach," *IEEE Sensors Journal*, Vol. 11, No. 11, 2011, pp. 3001–3008.

[41] Jha, S. K. and R. D. S. Yadava, "Denoising by Singular Value Decomposition and Its Application to Electronic Nose Data Processing," *IEEE Sensors Journal*, Vol. 11, No. 1, 2011, pp. 35–44.

Smart Grid

[42] Fadlullah, Z. Md., et al., "Toward Secure Targeted Broadcast in Smart Grid," *IEEE Communications Magazine*, Vol. 50, No. 5, May 2012, pp. 150–156.

[43] Lloret, J., P. Lorenz, and A. Jamalipour, "Communication Protocols and Algorithms for the Smart Grid," *IEEE Communications Magazine*, Vol. 50, No. 5, May 2012, pp. 126–127.

[44] Budka, K., et al., "GERI–Bell Labs Smart Grid Research Focus: Economic Modeling, Networking, and Security & Privacy," *1st IEEE Int. Conf. Smart Grid Communications (SmartGridComm)*, 2010, pp. 208–213.

[45] Sinha, A., S. Neogi, R. N. Lahiri, S. Chowdhury, S. P. Chowdhury, and N. Chakraborty, "Smart Grid Initiative for Power Distribution Utility in India," *IEEE Power and Energy Society General Meeting*, 2011, pp. 1–8.

[46] Hashmi, M., S. Hanninen, and K. Maki, "Survey of Smart Grid Concepts, Architectures, and technological Demonstrations Worldwide," *IEEE PES Conference on Innovative Smart Grid Technologies (ISGT Latin America)*, 2011, pp. 1–7.

[47] Güngör, V. C., et al., "Smart Grid Technologies: Communication Technologies and Standards," *IEEE Trans. Industrial Informatics*, Vol. 7, No. 4, November 2011, pp. 529–539.

[48] Vallejo, A., A. Zaballos, J. M. Selga, and J. Dalmau, "Next-generation QoS Control Architectures for Distributed Smart Grid Communication Networks," *IEEE Communications Magazine*, Vol. 50, No. 5, May 2012, pp. 128–134.

[49] Zhou, L., J. J. P. C. Rodrigues, and L. M. Oliveira, "QoE-driven Power Scheduling in Smart Grid: Architecture, Strategy, and Methodology," *IEEE Communications Magazine*, Vol. 50, No. 5, May 2012, pp. 136–141.

[50] Oksman, V. and J. Zhang, "G.HNEM: The New ITU-T Standard on Narrowband PLC Technology," *IEEE Communications Magazine*, Vol. 49, No. 12, December 2011, pp. 36–44.

[51] Liu, W., M. Sigle, and K. Dostert, "Channel Characterization and System Verification for Narrowband Power Line Communication in Smart Grid Applications," *IEEE Communications Magazine*, Vol. 49, No. 12, December 2011, pp. 28–35.

[52] Marmol, F. G., C. Sorge, O. Ugus, and G. M. Perez, "Do Not Snoop My habits: Preserving Privacy in the Smart grid," *IEEE Communications Magazine*, Vol. 50, No. 5, May 2012, pp. 166–172.

[53] Aragues, A., et al., "Trends in Entertainment, Home Automation and E-health: Toward Cross-Domain Integration," *IEEE Communications Magazine*, Vol. 50, No. 6, June 2012, pp. 160–167.

E-health

[54] Cova, G., et al., "A Perspective of State-of-the-Art Wireless Technologies for E-health Applications," *IEEE Int. Symposium on IT in Medicine & Education (ITIME)*, Vol. 1, 2009, pp. 76–81.

[55] Hairong, Y., X. Youzhi, and M. Gidlund, "Experimental E-health Applications in Wireless Sensor Networks," *WRI Int. Conf. on Communications and Mobile Computing* (CMC), Vol. 1, 2009, pp. 563–567.

[56] Tan, J., et al., "Gateway to Quality Living for the Elderly: Charting an Innovative Approach to Evidence-Based E-Health Technologies for Serving the Chronically Ill," *IEEE 13th Int. Conference on Computational Science and Engineering (CSE)*, 2010, pp. 146–159.

[57] Chowdhury, A., et al., "Radio over Fiber Technology for Next-Generation E-health in Converged Optical and Wireless Access Network," *Optical Fiber Communication Conference and Exposition (OFC/NFOEC) and the National Fiber Optic Engineers Conference*, 2011, pp. 1–3.

[58] Aragues, A., et al., "Trends and Challenges of the Emerging Technologies Toward Interoperability and Standardization in E-health Communications," *IEEE Communications Magazine*, Vol. 49, No. 11, November 2011, pp. 182–188.

[59] Agoulmine, N., P. Ray, and T.-H. Wu, "Efficient and Cost-Effective Communications in Ubiquitous Healthcare: Wireless Sensors, Devices and Solutions," (Guest Editorial), *IEEE Communications Magazine*, Vol. 50, No. 5, May 2012, pp. 90–91.

[60] Chowdhury, M. A., W. Mciver Jr., and J. Light, "Data Association in Remote Health Monitoring Systems," *IEEE Communications Magazine*, Vol. 50, No. 6, June 2012, pp. 144–149.

[61] Chan, V., P. Ray, and N. Parameswaran, "Mobile E-health Monitoring: An Agent-Based Approach," *IET Communications*, Vol. 2, No. 2, 2008, pp. 223–230.

[62] Corbellini, S., F. Ferraris, and M. Parvis, "A System for Monitoring Workers' Safety in an Unhealthy Environment by Means of Wearable Sensors," *IEEE Int. Instrumentation and Measurement Technology Conference (IMTC 2008)*, Victoria, Vancouver Island, Canada, 2008, May 12–15.

[63] Viswanathan, H., B. Chen, and D. Pompini, "Research Challenges in Computation, Communication, and Context Awareness for Ubiquitous Healthcare," *IEEE Communications Magazine*, Vol. 50, No. 5, May 2012, pp. 92–99.

[64] Caldeira, J. M. L. P., J. J. P. C. Rodrigues, and P. Lorenz, "Toward Ubiquitous Mobility Solutions for Body Sensor Networks on Healthcare," *IEEE Communications Magazine*, Vol. 50, No. 5, May 2012, pp. 108–115.

[65] Mitra, U., et al., "KNOWME: A Case Study in Wireless Body Area Sensor Network Design," *IEEE Communications Magazine*, Vol. 50, No. 5, May 2012, pp. 116–125.

[66] Rahman, M. A., M. F. Alhamid, W. Gueaieb, and A. El Saddik, "An Ambient Intelligent Body Sensor Network for E-health Applications," *IEEE Int. Workshop on Medical Measurements and Applications* (MeMeA), 2009, pp. 22–25.

[67] Barakah, D. M. and M. Ammaduddin, "A Survey of Challenges and Applications of Wireless Body Area Network (WBAN) and Role of a Virtual Doctor Server in Existing Architecture," *Third Int. Conference on Intelligent Systems, Modelling and Simulation (ISMS)*, 2012, pp. 214–219.

[68] Ullah, S., et al., "A Comprehensive Survey of Wireless Body Area Networks," *J. Med. Syst.*, Springer, August 2010, pp. 1–30.

[69] Boulis, A., D. Smith, D. Miniutti, L. Libman, and Y. Tselishchev, "Challenges in Body Area Networks for Healthcare," *IEEE Communications Magazine*, Vol. 50, No. 5, May 2012, pp. 100–106.

[70] Noimanee, K., et al., "Development of E-health Application for Medical Center in National Broadband Project," *Biomedical Engineering International Conference (BME-iCON)*, 2011, pp. 262–265.

[71] Nita, L., M. Cretu, and A. Hariton, "System for Remote Patient Monitoring and Data Collection with Applicability on E-health Applications," *7th Int. Symposium Advanced Topics in Electrical Engineering (ATEE)*, 2011, pp. 1–4.

[72] Ying, S., and J. Soar, "Integration of VSAT with WiMAX Technology for E-health in Chinese Rural Areas," *2010 Int. Symp. Computer Communication Control and Automation (3CA)*, Vol. 1, pp. 454–457.

[73] ISO/IEEE11073 — Personal Health Devices Standard (X73PHD), Health Informatics [P11073-00103, tech. rep., overview] [P11073-104zz. Device Specializations] [P11073-20601, Application Profile — Optimized Exchange Protocol], http://standards.ieee.org.

[74] Rahman, M. A., W. Gueaieb, and A.El Saddik, "Ubiquitous Social Network Stack for E-health Applications," *IEEE International Workshop on Medical Measurements and Applications Proceedings (MeMeA)*, 2010, pp. 57–62.

Traffic control

[75] Federal Communications Commission (FCC) Fact Sheet, "FCC Wireless 911 Requirements", 2001.

[76] EU Institutions Press Release, "Commission Pushes for Rapid Deployment of Location Enhanced 112 Emergency Services," DN:IP/03/1122, Brussels, Belgium, July 2003.

[77] Hartenstein, H. and K. P. Laberteaux, "A Tutorial Survey on Vehicular Ad Hoc Networks," *IEEE Communications Magazine*, Vol. 46, No. 6, June 2008, pp. 164–171.

[78] Karagiannis, G., et al., "Vehicular Networking: A Survey and Tutorial on Requirements, Architectures, Challenges, Standards and Solutions," *IEEE Communications Surveys & Tutorials*, Vol. 13, No. 4, Fourth Quarter 2011, pp. 584–616.

[79] Sichitiu, M. L. and M. Kihl, "Inter-Vehicle Communication Systems: A Survey," *IEEE Communications Surveys & Tutorials*, Vol. 10, No. 2, 2nd Quarter 2008, pp. 88–105.

[80] Suthaputchakun, C. and Z. Sun, "Routing Protocol in Intervehicle Communication Systems: A Survey," *IEEE Communications Magazine*, Vol. 49, No. 12, December 2011, pp. 150–156.

[81] Acosta-Marum, G. and M. A. Ingram, "Six Time- and Frequency-Selective Empirical Channel Models for Vehicular Wireless LANs," *IEEE Vehicular Technology Magazine*, Vol. 2, No. 4, December 2007, pp. 4–11.

[82] Molisch, A. F., F. Tufvesson, J. Karedal, and C. F. Mecklenbrauker, "A Survey on Vehicle-to-Vehicle Propagation Channels," *IEEE Wireless Communications*, Vol. 16, No. 6, December 2009, pp. 12–22.

[83] de Vogel, G. F., P. K. Baccei, and P. T. Shaw, "The United States Navy Navigating in the 21st Century," *MTS/IEEE Conf. and Exhibition, Oceans'2001*, Vol. 3, pp. 1460–1465.

Biography

Mehmet Şafak received the B.Sc. degree in Electrical Engineering from Middle East Technical University, Ankara, Turkey in 1970 and M.Sc. and Ph.D. degrees from Louvain University, Belgium in 1972 and 1975, respectively.

He joined the Department of Electrical and Electronics Engineering of Hacettepe University, Ankara, Turkey in 1975. He was a postdoctoral research fellow during the academic year 1975–1976 in Eindhoven University of Technology, The Netherlands. From 1984 to 1992, he was with the Satellite Communications Division of NATO C3 Agency (formerly SHAPE Technical Centre), The Hague, The Netherlands, as a principal scientist. During this period, he was involved with various aspects of military SATCOM systems and represented NATO C3 Agency in various NATO committees and meetings. In 1993, he joined the Department of Electrical and Electronics Engineering of Eastern Mediterranean University, North Cyprus, as a full professor and was the Chairman from October 1994 to March 1996. Since March 1996, he is with the Department of Electrical and Electronics Engineering of Hacettepe University, Ankara, Turkey, where he acted as the Department Chairman during 1998–2001. He is currently the Head of the Telecommunications Group.

He conducted and supervised projects, served as a consultant and organized courses for various companies and institutions on diverse civilian and military communication systems. He served as a member of the executive committee of TUBITAK (Turkish Scientific and Technical Research Council)'s group on electrical and electronics engineering and informatics. He acted as reviewer in various national and EU projects and for distinguished journals. He was involved in the technical programme committee of many national and international conferences. He served as the Chair of 19th IEEE Conference on Signal Processing and Communications Applications (SIU 2011). He represented Turkey to COST Action 262 on Spread Spectrum Systems and Techniques in Wired and Wireless Communications. He acted as the chairman of the COST Action 289 Spectrum and Power Efficient Broadband Communications.

He was involved with high frequency asymptotic techniques, reflector antennas, wave propagation in disturbed SATCOM links, design and analysis of military SATCOM systems and spread spectrum communications. His recent research interests include multi-carrier communications, channel modelling, cooperative communications, cognitive radio and MIMO systems.

8

Green Wireless Sensor Networks with Distributed Beamforming and Optimal Number of Sensor Nodes

H. Nikookar

Delft University of Technology, The Netherlands

8.1 Introduction

The proliferation of wireless technology and the advancement in power and size-efficient computing has allowed the use of small radio sensor devices to be employed in remote areas for different applications such as disaster management, combat field reconnaissance, border protection and surveillance. The devices in such applications are usually deployed in large numbers in unattended environments to sense or gather useful information. For a variety of reasons the devices are deployed at random locations without prior knowledge of the topology of the network. Related with their small physical size, these devices are provided with low memory, low data rates, low processing power and limited energy supply (i.e., they possess tiny low-power batteries and small omni-directional antennas). Since it is unlikely that these wireless sensor devices be periodically monitored for the battery replacement, efficient

COmmunications- NAvigation-SENsing-SErvices (CONASENSE), 165–179.

power consumption of these devices is critical. However, the sensor devices need to report the gathered information back to a base station also known as a sink node which is usually located in ranges of kilometers away from the radio sensor network. A single device cannot sustain to form a communication link with the sink node out of its battery power supply. Furthermore, there is no way that each device can focus its transmit signal in the direction of the sink since they have only omni-directional antennas. For these reasons there are limitations in each device in terms of required transmit power and directivity. This is where distributed beamforming (DB) in future wireless sensor networks, comes into the picture. The individual wireless sensor devices are used as virtual antenna elements which cooperatively send the same transmit signal simultaneously to improve the range and directivity. DB is a cooperative scheme applied to improve range and directivity of sensor devices in a wireless radio environment. DB plays a double benefit in the cognitive radio sensor networks. The improved directivity of the DB technique allows to focus energy not only to boost power in the desired direction but also is used to minimize and reduce power in all undesired directions. However, efficient use of the DB technique requires a number of practical issues that need to be taken into consideration. Among them is the monitoring of the optimal number of wireless nodes to be used in DB so that the overall consumption is reduced. This ensures to maintain the "green" aspect of the network. The available selection schemes for the optimal number of nodes in DB techniques for wireless sensor networks do not address the cognitive radio scenarios where minimization of interference in the direction of primary users (PU) is of utmost importance. In this chapter a selection method with a close to optimal number of nodes for DB in cognitive radio network is proposed. Simulation results show a saving of 85% of the total required energy per transmission, an amount which remarkably enhances the greenness of the radio sensor network. The rest of this chapter is organized as follows. In Section 2 DB in wireless sensor networks is reviewed. In Section 3, we illustrate the total energy consumption of the network and quantify the energy consumed per single transmission. The sub-optimality of interference minimization is also presented. In Section 4, the use of the *clustering* node selection method for operating close to the optimal number of nodes is discussed. In Sections 5 and 6, simulation results and the conclusion are given, respectively.

8.2 Distributed Beamforming in Wireless Sensor Networks

With the rapid advancements of wireless technology, the use of wireless sensor nodes has increased for a wide variety of applications. Distributed beamforming (DB) as a cooperative scheme plays a major role in harnessing the limitation in transmit power of the individual sensor devices. The sensor devices are usually battery driven and are deployed in remote areas where periodic battery replacement is unlikely [1]. Therefore, these devices have to rely on their battery power supply for a long period of time.

In DB, wireless sensor nodes minimize the energy spent per node by cooperatively transmitting the same signal simultaneously so that the signal from each node is added constructively at the receiver. Thus, the gain of the distributed array increases with the number of nodes [2]. This further means that the transmit signal power per node decreases with increasing number of nodes for a fixed bit energy to noise power spectral density ratio at the receiver i.e., $\frac{E_b}{N_0}$.

However, prior to transmission the beamforming nodes have to share information in a coordinated way so that all nodes can transmit the same signal. The information exchange during the pre-transmission phase creates increasing overhead in terms of energy consumption as more nodes are used in beamforming.

As a result, the energy consumption of the network increases. Thus, it is critical to choose the number of nodes to be used in beamforming to optimize the total energy consumption of the network and thus maintaining greenness in wireless sensor network (WSN).

The authors in [3] have provided a practical framework for optimal number of nodes to be used in a single-hop WSN where every node acts as a relay in the inter-node communication within the sensor network. They have calculated the optimal number of nodes from the consideration of the total energy consumption of the network only.

In the cognitive radio (CR) sensor network where the utilized spectrum of PU allows for cognitive radio nodes access, it is required to limit the transmit power toward the PU. The PU usually set an interference limit P_0 which is the maximum power level that CR users can transmit in the direction of PU. The power transmitted in the direction of PU can be reduced by increasing the number of nodes according to the inverse relationship between the array sidelobe level and the number of nodes, which form the array pattern of the CR network.

Therefore, the number of nodes to be used in beamforming should be large enough to reduce the power transmitted toward PU and push it below a given P_0. As a result, the number of nodes to be used in DB for CR will not be optimal but sub-optimal taking into account the interference minimization schemes and extra energy is spent to reduce interference in the direction of PU.

In this chapter, a new node selection method is presented that leads to operating close to the optimal number of nodes through reduced sidelobe level of the beam pattern in the direction of PU allowing the minimization of interference with less number of nodes. Therefore, by using this method the number of nodes is reduced leading to energy conservation which can be quantified in the order of Joules and hence maintaining the greenness of the network.

8.3   Optimizing Energy Consumption of Cognitive WSN

8.3.1   Quantifying Energy Consumption per Transmission

In order to quantify the energy consumption of the network of N nodes, we categorize a single transmission in two stages, i.e., the pre-transmission phase and transmission phase. The wireless sensor nodes are uniformly distributed in a circular disk of radius Z as shown in Figure 4.1. The node which is selected to coordinate inter-node communication, known as the Cluster Head (CH), is located at the center of the circle and has location information of all nodes. The transmit power of the nodes suffices to form a communication link with the CH. We further assume that CH also participates in the beamforming process.

A sensor node first sends its transmit information to the CH and the following coordination happens during the *pre-transmission phase* in order to facilitate signal transmission to a far field CR receiver:

> **Step 1:** A cluster head decides the number of nodes N to be used in beamforming and sends participation requests to $N - 1$ nodes.
> **Step 2**: $N - 1$ nodes confirm their participation by sending out their reply to the CH
> **Step 3**: The cluster head sends the transmit signal to the selected $N - 1$ nodes along with the transmit power level they should be

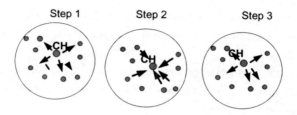

Fig. 4.1a Pre-transmission Phase.

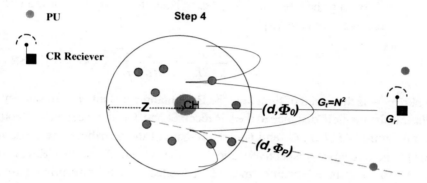

Fig. 4.1b Transmission Phase.

using. The cluster head can calculate the transmit power level of each node for a given destination and the desired $\frac{E_b}{N_0}$ value at CR receiver as will be done later in this section.

Step 4: Then N nodes, including the CH itself, transmit the signal simultaneously to a far field CR receiver during the transmission phase

The total energy consumption can be quantified using the framework provided in [3] for a single hop transmission which we adopted to our transmission scheme in which CH is reachable by every node while assuming the following assumptions:

— The number of nodes is optimized for a fixed bit energy to noise ratio $\frac{E_b}{N_0}$ at the CR receiver.
— Every node transmits at a fixed rate of R bits/sec a set of symbols: S_1 symbols during the pre-transmission phase and S_2 symbols during transmission phase.

— The nodes use a transmit power level p_1 during the *pre-transmission phase* and p_2 during signal transmission to a far field CR receiver.

— A node consumes a power level of *a Watts* when its transmitter is activated and *b Watts* when its receiver is activated during the reception of a signal.

— The distributed network of N nodes uses a total transmit power P_t during the *transmission phase*. Thus, for a fixed $\frac{E_b}{N_0}$ at the CR receiver, we have [3]:

$$\frac{E_b}{N_0} = \frac{P_t G_t G_r}{kTRL_0} \left(\frac{c}{4\pi d f} \right)^2 \tag{8.1}$$

where c — is the speed of light, k is the Boltzmann constant, f is frequency in Hz, d is the far-field distance to the CR receiver and T is the receiver's effective temperature in *Kelvin*. G_t and G_r are the gain of the distributed network and the far-field receiver, respectively and L_0 is the system loss at the CR receiver.

For a distributed beamformer of N nodes, the beamforming gain is $G_t = N^2$ [2]. Thus, Equation (8.1) can be rewritten as:

$$P_t = \frac{1}{N^2} \left[\frac{E_b}{N_0} \frac{kTRL_0}{G_r} \left(\frac{4\pi d f}{c} \right)^2 \right] \tag{8.2}$$

Therefore, the transmit power p_2 per node during the transmission phase equals:

$$P_2 = \frac{P_t}{N} = \frac{1}{N^3} \left[\left(\frac{E_b}{N_0} \right) \left(\frac{kTRL_0}{G_r} \right) \left(\frac{4\pi d f}{c} \right)^2 \right] \tag{8.3}$$

The total energy consumption of the network per transmission is the sum of the total energy consumed during the *pre-transmission phase* and *transmission phase* which is quantified below in Joules.

Pre-transmission:

Step 1: $(N - 1) \left(\frac{a+b+p_1}{R} \right) S_1$

Step 2: $(N - 1) \left(\frac{a+b+p_1}{R} \right) S_1$

Step 3: $(N - 1) \left(\frac{a+b+p_1}{R} \right) S_2$

Transmission:

Step 4: $N\left(\frac{a+p_2}{R}\right)S_2$

Therefore, the total energy consumption in Joules can be found by adding the energy consumed both in the *pre-transmission* and *transmission phase* and substituting in Equation (8.3) for P_2, we get:

$$E = N\left(\frac{a+b+p_1}{R}\right)(2S_1 + S_2) + NS_2\frac{a}{R}$$

$$+ \frac{S_2}{N^2}\left(\frac{E_b}{N_0}\right)\left(\frac{4\pi df}{c}\right)^2\left(\frac{kTL_0}{G_r}\right) - \left(\frac{a+b+p_1}{R}\right)(2S_1 + S_2) \quad (8.4)$$

8.3.2 Energy and Interference Minimization

When more nodes are used, the energy consumed during the pre-transmission phase (steps 1–3) increases while the energy consumed during the transmission phase i.e., step 4 decreases as shown in Figure 8.2. The optimal number of nodes N^* is determined by minimization of Equation (8.4). The total energy consumption is optimized for the number of nodes and a fixed $\frac{E_b}{N_0}$ value at the CR receiver which in this case equals 17 dB. It is shown in Figure 8.2 that the

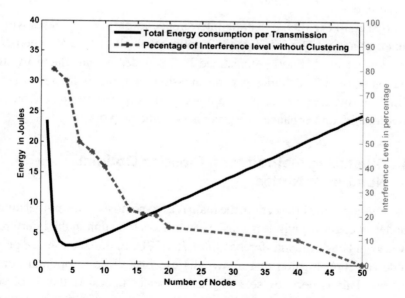

Fig. 8.2 Energy consumption and interference minimization versus number of nodes.

optimal number of nodes becomes $N^* = 5$ and the total energy consumed is only 2.95 Joules.

However, simulation results show that the optimal number of nodes is not enough to achieve a sufficient reduced sidelobe level in the direction of PU. This was verified by monitoring the level of interference generated in the direction of PU according to the cumulative distribution function (CDF) of the beam pattern. The CDF was generated using Monte Carlo simulations of 1000 iterations.

We scaled the interference generated according to the percentage of the time the power level in the direction of PU exceeded the interference limit P_0. If the power transmitted in the direction of PU is greater than P_0 for I_0 % of the time, we denote the interference level as I_0 %.

However, the interference level for $P_0 = 0.1$ Watt (20 dBm) produces significant interference as large as 70% when only the optimal number of $N^* = 5$ nodes are used. The use of more nodes brings a reduced sidelobe level and hence less interference toward PU. Therefore, it becomes compelling to pay a price by using more nodes, i.e., a sub-optimal number N_1, in order not to exceed the interference limit of PU. This makes the task of interference minimization energy demanding as more energy is consumed through the use of more nodes.

It is seen in Figure 8.2 that the interference level in % decreases with the number of nodes. The interference level reaches close to 0 % when $N_1 = 50$ nodes are used. Energy consumed by 50 nodes as calculated by using Equation (8.4) is 24.6 Joules per transmission, and thus is the result is sub-optimal. In this chapter, we present a node selection method that avoids this sub-optimality and enhances the greenness of the network.

8.4   Clustering Method for a Close to Optimal Number of Nodes

As mentioned in the last section, the main reason not to use the optimal number of nodes in cognitive wireless sensor networks is to maintain the interference power level below the interference limit P_0 of PU. In this section, we present a node selection method that allows the use of a close to optimal number of nodes. This is done by selecting a number of nodes M that is close to the optimal number N^*. By using M nodes, combined with a selection of clustering nodes, the power transmitted in the direction of PU is kept below

P_0 close to 100% of time. Therefore, through node clustering we save $N_1 - M$ nodes from participating in the beamforming. This also means that a significant amount of energy consumption is reduced, meaning that the greenness of the network is improved.

8.4.1 Selection Method for Node Clustering

Node selection is done centrally at the *CH*. The *CH* starts calculating the optimal number of nodes. We assume that the nodes are distributed uniformly in a circular disk of radius Z and the far field distance to PU from each node is denoted as d_k. Then *CH* calculates Ω_k, i.e., the far field phase of the transmit signal from all nodes using the location information and the direction of PU. The calculated phase is used as an input for the *clustering* process. The expression for Ω_k becomes

$$\Omega_k = \Psi_k + \frac{2\pi}{\lambda} d_k(\Phi_p) \tag{8.5}$$

The first term in Equation (8.5) is the initial phase of each node which is a function of the far field distance and the direction Φ_0 of the intended CR receiver as given in [1]. The second term is the propagation phase of the transmit signal at the far field in the direction Φ_P of primary user, where λ is wavelength of the transmit signal. The calculated phase for each node Ω_k is used by the *k-means* clustering algorithm as a criterion to group every node under a single phase group, also referred to as a cluster. The implementation of the *k-means* clustering algorithm is given in [4]. The CH continues by picking a single node from each single phase group (*cluster*) until the intended number of nodes is selected.

As a result, the nodes selected have a different phase from each other allowing for destructive addition of the signals in order to obtain a reduced sidelobe level in the direction of PU. The CH then monitors the interference level in the direction of PU by calculating the CDF of the beam pattern.

If the interference generated is below P_0 for almost 100% of the time, then the optimal numbers of nodes are used for beamforming. Otherwise, the number of nodes is increased by one until the interference generated is below the interference limit P_0. Depending on the value of P_0, the *clustering* method may converge to an optimal number of nodes or close to optimal. Nonetheless, the inherent destructive addition of the far field phases of the selected nodes in the direction of PU guarantees us to an operate close to optimal number of

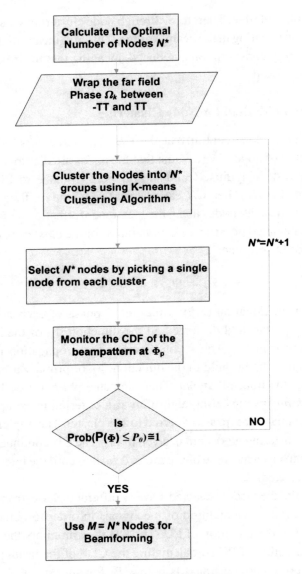

Fig. 8.3 *Clustering* node selection method.

nodes. The *clustering* node selection method is summarized in Figure 8.3. The sidelobe level reduction obtained using the proposed *clustering* node selection method is shown in Figure 8.4 for a PU located at $\Phi_P = 20^0$. The reduced

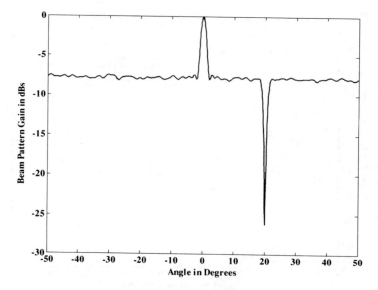

Fig. 8.4 Average beam pattern of 6 nodes.

sidelobe level allows interference minimization. Thus, for a given value P_0, the *clustering* technique can be used to address the interference avoidance with less number of nodes allowing the use of a close to optimal number of nodes for beamforming. The use of less number of nodes enhances the greenness concept of the network as demonstrated in the simulation section.

8.5 Simulation Results

We simulated a group of nodes that are distributed uniformly in a disk of a circular area with a radius $Z = 32$ meters. A single node is located in an area of 3.21 m^2. The beamforming nodes transmit at a bit rate of $R = 500$ Kbps. The intended CR receiver is located $d = 40$ km away from the group of sensor nodes. The energy of both the CR transmitter and receiver are $a = 100$ μW, $b = 100$ μW respectively; the nodes use a transmit power $p_1 = 10$ mW for inter-node communication. The transmit signal per single transmission is $S_1 = 2.5$ Mbits long and the inter-communication signal $S_2 = 1$ Kbits long. The far-field receiver has omni-directional antenna i.e., $G_r = 1$. The beamforming nodes have a transmit gain $G_t = N^2$. The interference limit is $P_0 = 20$ dBm. The bit energy per noise spectrum ratio at the receiver is $\frac{E_b}{N_0} = 17$ dB.

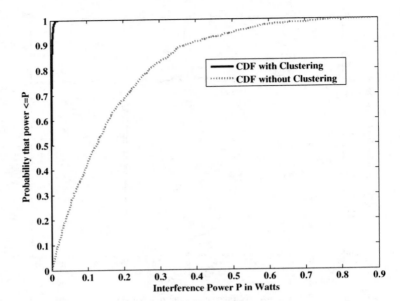

Fig. 8.5 Comparison of CDF of the beam pattern at $\Phi_P = 20°$ with and without node selection for 6 nodes.

The optimal number of nodes that minimizes energy consumption of the network for the above simulation scenario calculated by minimization of Equation (8.4) is $N^* = 5$. Our *clustering* node selection method converged to $M = 6$ nodes. Through comparison of the CDF of the beam pattern with and without *clustering* method, we were able to verify that we can minimize interference with less number of nodes than normally is required. The CDF comparison is shown in Figure 8.5 when 6 nodes are used for beamforming. For $P_0 = 20$ dBm, the power transmitted using the *clustering* method is less than P_0 for close to 100% of the time. Hence, the *clustering* node selection method can avoid interference with 6 nodes which is close to the optimal number 5.

It is shown in Figure 8.6 that the interference power can be kept below the interference limit with the use of sub-optimal $N_1 = 50$ nodes without *clustering* which consume 24.6 Joules of energy per transmission. However, it can be done also with the use of only 6 nodes with *clustering*. Therefore, using the proposed node selection method, 44 nodes have been saved from participating in beamforming and this brings energy reduction of about 21 Joules

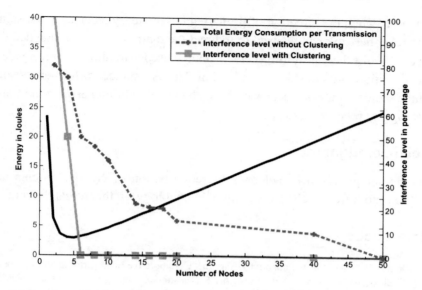

Fig. 8.6 Interference level and energy consumption vs. number of nodes for a threshold power level $P_0 = 20$ dBm.

per transmission which is 85% of the energy required in the method without clustering.

8.6 Conclusion and Further Research

Minimizing the total energy consumption per single transmission is critical for maintaining the greenness of wireless sensor networks. Optimizing the energy consumption provides the optimal number of nodes in WSN. However, for CR, the sub-optimality of the energy consumption is inevitable due to extra strict requirement of power transmitted toward PU. The clustering node selection technique suggested in this research work solves the problem by allowing beamforming with a close to optimal number of nodes to be selected while keeping the interference to PU below the interference limit. In this study, no channel conditions were taken into consideration in the analysis. Looking into multipath and shadowing effects will bring distributed beamforming one step further into the practical deployment. Furthermore, here the node selection was used for transmit distributed beamforming without consideration of relay networks. The use of a relay can be employed in the cognitive radio sensor networks when the direction of primary and secondary receiver is the same. The relay network can be used to redirect the main beam to an intended CR receiver while the transmitting network can be used to minimize interference in that direction through sidelobe reduction. This is a subject of further research which requires the consideration of the channel condition between the transmitting and relaying network.

Acknowledgment

The author greatly acknowledges the contribution of N. M. Tessema and X. Lian (both from Delft University of Technology) in this research work.

References

[1] Ochiai, H., P. Mitran, V. Poor, and V. Tarokh, "Collaborative Beamforming for Distributed Wireless Ad hoc Sensor Networks", *IEEE Transactions on Signal Processing*, Vol. 53, No. 11, pp. 4110–4124, Nov. 2005.

[2] Uher, J., T. A. Wysocki, and B. J. Wysocki, "Review of Distributed Beamforming", *Journal of Telecommunications and Information Technology 2011*, USA, Jan 2011.

[3] Poor, V., M. Tummala, and J. McEachen, "Optimizing the Size of an Antenna Array", *Fortieth Asilomar Conference on Signals, Systems and Computers, 2006, ACSSC'06*, pp. 2281–2284, Oct. 2006–Nov. 1 2006.

[4] Kanungo. T., D. Mount, N. Netanyahu, D. Piatko, C. Silverman, and Y. Wu, "An Efficient K-means Clustering Algorithm: Analysis and implementation," *IEEE Transactions on Pattern Analysis and Machine Intelligence*, Vol. 24, No. 7, pp. 881–892, Jul. 2002.

[5] Lian, X., H. Nikookar, and L. P. Ligthart, "Distributed Beamforming with Phase-only Control for Green Cognitive Radio Network", *Eurasip, Journal of Wireless Communication and Networking*, Vol. 65, Feb. 2012.

[6] Ahmed, M. F. A. and S. A. Vorobyov, "Sidelobe Control in Collaborative Beamforming via Node Selection," *IEEE Trans on Signal Processing*, Vol. 58, No. 12, pp. 6168–6180, Dec. 2010.

[7] Tessema, N. M., X. Lian, and H. Nikookar, "Efficient Node Selection Techniques in Green Cognitive Radio Networks", *European Microwave Week 2012*, Amsterdam, Oct. 2012.

Biography

Homayoun Nikookar is an Associate Professor in the Faculty of Electrical Engineering, Mathematics and Computer Science at Delft University of Technology where he leads the Radio Advanced Technologies and Systems (RATS) program and supervises a team of researchers carrying out cutting-edge research in the field of advanced radio transmission. He has received several paper awards at international conferences and symposiums and the 'Supervisor of the Year Award' at Delft University of Technology in 2010. He is the Secretary of the new scientific society on Communication, Navigation, Sensing and Services (CONASENSE). He has published more than 130 refereed journal and conference papers, coauthored a textbook on 'Introduction to Ultra Wideband for Wireless Communications, Springer 2009', and has authored the book 'Wavelet Radio, Cambridge University Press, 2013'.

9

Battle on Frequency Spectrum Use for Radio Navigation and Radio Communications: It Cannot Go Forever — We Have to Find the Solution!*

Durk van Willigen

Delft University of Technology, em. Prof. Dr Reelektronika BV, CEO, The Netherlands

9.1 Introduction

The discussions in the United Stated between LightSquared, the FCC, the GPS industry, DoD, DoT and users about the division of spectrum in the L-band should not be seen as just an internal US affair. The strength of both camps, Telecom providers and the GPS industry, may indicate that this battle may easily expand to other parts of the world. For many, this battle was a surprise emerging in the generally peaceful navigation world where discussions are more gentlemen-like and mostly focused on which GNSS systems is the best, or on the backup of GNSS which is so vulnerable and where society cannot function anymore without GNSS. All this changed abruptly when LightSquared published plans to install 40,000 transmitters in a band adjacent to the L1 band used by GPS. This led to numerous protests of the GPS industry and users what has been published in many magazines. So, is it Telecom versus GPS, or is a sensible cooperation in reach?

---

*The paper was published in Coordinates (Vol. VIII, Issue 1, January 2012).

COmmunications- NAvigation-SENsing-SErvices (CONASENSE), 181–190.

Many may wonder whether these debates are really important and why this exploded so unexpectedly. A look back into radio navigation history may help to understand the underlying discomfort on both sides. Radio navigation started as terrestrial based systems, and some 30 years ago space-based systems took over.

9.2   Review on Radio Navigation

9.2.1   Omega

The first worldwide radio navigation system was Omega, which basically used four carrier wave signals in the 10–14 kHz band. For identification and wide-laning purposes, the four carriers were switched on and off in an accurately defined pattern. The radiated signals propagated quit easily in the layer between the earth surface and the ionosphere. The attenuation is those layers is moderate, so, the 10 kW EIRP signals could be received worldwide. As the propagation models are complex and not accurately known, the attainable accuracy was in the range of some kilometres. As VLF signals can also be received under water, the navy's interest at that time was understandable. However, the overwhelming introduction of GPS made Omega disappear rapidly. However, it is important to remember that the on-off modulation pattern of Omega required a total spectrum bandwidth of approximately 160 Hz.

9.2.2   Loran-C

Loran-C, operating at 100 kHz, showed a much better accuracy. Although the system was originally specified to achieve absolute accuracy levels of better than one quarter mile, 463 metres, in practice far much better results were obtained. The main reason was a better understanding of the propagation phenomena, and the possibility to discriminate between groundwave and sky-wave signals. This is achieved by applying pulse-like amplitude modulation instead of using carrier waves only. As the skywave path is longer than that of the groundwave, the receiver can relatively easy select the groundwave for the position determination. This ground/sky wave discrimination works well up to distances of 1,000 km or more. The models of the groundwave propagation are much more accurately known than that of Omega which results in accuracy levels down to 50 metres. If differential Loran-C techniques are applied, absolute accuracies of better than 10 metres are attainable. Loran-C, and its successor eLoran, is in use in many parts of the world like northwest Europe, Russia (called Chayka), South Korea, Japan, India and the Middle East. To overcome the high atmospheric noise levels in the 100 kHz band, high-power transmitters in the range of 100 kW to 2 MW were a necessity. For a long time these high energy levels were considered as an economic disadvantage, but today it is a blessing as jamming these strong signals over larger areas is very difficult. This makes eLoran a very capable and efficient backup for today's GNSS systems which are easily denied over rather large areas with simple low-power jammers. The Loran pulse modulation of the 100 kHz carrier implies a larger claim on the spectral bandwidth and amounts to 20 kHz.

9.2.3 DECCA

DECCA is another LF system which also has been decommissioned after GPS became operational. This system used a number of carriers which were on-off modulated in well-defined rhythms' as identification of the station. The required total spectral bandwidth was about 120 Hz. Due to the carrier-wave type of signals, ground-waves cannot be separated from sky-waves which, depending on the ionosphere conditions, limited the working range down to a 100 km which made it a typical coastal navigation system. The accuracy was quite good and could be around 50 metres.

9.2.4 Transit and Tsiklon

When the Russians launched the first Sputnik satellite, the Americans listened to the broadcast signals and observed a rather strong Doppler shift of the received signals. These Doppler shifts made computation of the satellite's orbit feasible. Then US scientists developed the first space-based navigation system called Transit, and Tsiklon in the former USSR, which worked in an opposite manor. The orbits of the satellites are now accurately known and by measuring the Doppler shifts over some time, the user position could be established. Accuracies were in the order of 10–50 metres, and the systems could be used worldwide. Transit broadcasts two carriers on 150 and 400 MHz, respectively. The Doppler peak-to-peak shift ranged from 4 kHz on 150 MHz to 10 kHz on 400 MHz. So, the total spectrum bandwidth was a mere 14 kHz.

9.2.5   GPS and Glonas

The real major step in all aspects of radio navigation was done by GPS in the US and GLONASS in Russia. These systems applied CDMA code for ranging and data transfer and the signals are spread-spectrum modulated on the L1, L2, and L5 bands. Although the accuracy of these worldwide systems is an impressive 2–20 m, the spectrum requirements are also impressive being approximately 60 MHz. This bandwidth is primarily needed for ranging which is based on correlation of the received signals with the replica code of the selected satellite. The data transmissions need less than 1 kHz bandwidth.

9.3   Frequency Spectrum Use of Present and Future Radio Navigation Systems

The table shows the enormous differences in spectrum needs for a number of well-known systems.

	System	Spectral requirement
Terrestrial based	Omega	160 Hz in VLF-band
	DECCA	120 Hz LF-band
	Loran-C/eLoran/Chayka	20 kHz in LF-band
Space based	Transit	15 kHz in VHF-band
	GPS/GLONASS/Compass/Galileo	60 MHz in L-band

This leaves us with two burning questions whether these enormous spectrum claims by GNSS are really needed, and were at the start of designing GPS, alternative solutions considered that have more modest bandwidth claims without losing performance? Some background is presented in next section.

9.4   Bandwidth Requirements in Radio Navigation

9.4.1   Ranging Based on CDMA

CDMA offers many advantages in respect of accurate measuring the time of arrival of the signals, rejecting multipath, simple decoding of the received signals, and also that all satellites can broadcast in the same part of the spectrum. For the military it was important that these signals could not be used when the code was not known. Further, the applied spread-spectrum technique makes it difficult to detect the unknown received signals which are some 20 dB below the galactic noise level. The disadvantage of CDMA, however, is that range measurements require a rather large bandwidth to obtain sharp correlation responses, a prerequisite for accurate ranging.

9.4.2 Ranging Based on Carrier Tracking

It is interesting to see how Omega was designed. All eight stations broadcast sequentially four different carriers in the VLF band. On basis of the way the frequency steps were formatted, the receiver could identify which station was broadcasting on one of the four frequencies. Although the accuracy was rather poor, the basic concept was quite ingenious as it offered worldwide a simple carrier tracking ranging technique with a total bandwidth of just a mere 160 Hz.

9.4.3 Theoretical Exercise: GNSS Based on Carrier Tracking Instead of CDMA

Although GNSS signal concept cannot be changed anymore, it is still interesting to see whether a pure carrier-tracking technique could also lead to an accurate GNSS navigation system with a large reduction in spectral bandwidth. For example, assume that a satellite sends on L1, L2 and L5 a carrier, each with two sub-carriers. This would result in 9 carriers in total. Further, we assume that only carrier tracking is used for range measurements in order to keep the bandwidth limited. The largest unambiguous range should be about 30,000 km for MEO satellites. This can be achieved by using two carriers separated by 10 Hz. The phase difference between the two carriers which started at the same time in the SV is a measure of the distance between de SV and the receiver. Generally speaking, we assure that this range measurement can be done with an accuracy of better than 10% of the unambiguous range, so 3,000 km in this example. The next step is then to use two frequencies at 100 Hz apart, which would yield an unambiguous range of this 3,000 km and with an accuracy of 300 km. In practise, the precision on L-band frequencies is better than 10%, more around 1% of the wavelength. If we would use the following set of nine carrier frequencies, we would end up in carrier tracking without ambiguous range problems down to a precision at L1 of a few mm. By adding quadrature modulation on for example L1 and L2, orbital and time information of each SV can be received by the user.

Frequency	Unambiguous range	Precision 1% of wavelength
L1	0.2 m	2 mm
L1 + 10 Hz	30,000 km	300 km
L1 + 100 Hz	3,000 km	30 km
L2 + 1.5 kHz	200 km	2 km
L2 + 20 kHz	15 km	150 m
L5 + 300 kHz	1 km	10 m
L5 + 5 MHz	60 m	0.6 m
L1 − L2 => 350 MHz	0.86 m	8.6 mm
L2 − L5 => 50 MHz	6 m	60 mm

Ionosphere data can be retrieved from the differences between de basic carriers on L1, L2 and L5. This relatively simple concept is modest in its spectrum needs. Although only carriers are used, Doppler effects will consume 30 kHz on L1, 24 kHz on L2 and 21 kHz on L5, totalling to 75 kHz. The data modulation is just a small fraction of the Doppler. Further, as all satellites would operate on a different frequency, no mutual interference would be experienced. So, for one hundred SVs, the total spectrum needed is limited to 7.5 MHz. Two additional advantages of this concept are that although more satellites will consume more spectrum, this will not lead to an increased noise floor as is the case with the CDMA structure of GNSS. Due to the applied carrier technique it would be more difficult to jam; the power density is much higher, and the receiver tracking bandwidth is rather small. Jamming over the entire spectral bandwidth would require significantly much more power than with CDMA techniques.

However, the above given idea is just an exercise as modifying the current GNSS structures is out of the question. But the present conflict between Light-Squared and GPS would be easier to solve if we would have such a spectrum-efficient system.

9.5 Conflict between LightSquared and GPS

The basic issue between LightSquared is the so called comfort-zone, the separation between the GPS spectrum and that of LightSquared. GPS receivers cannot apply brick-stone type of bandpass filters in front of the LNA in the antenna without sacrificing its noise performance and still be able to show excessive attenuation for LightSquared signals. Although JAVAD published that they have developed a useful solution, we have not seen yet objective tests to confirm their claim. The large bandwidth need of GPS is easier to understand by looking at the eye-pattern of GPS and digital telecom signals. GPS receivers need very steep slopes of the eyes in order to make sharp correlation peaks feasible, while telecom receivers just need to discriminate between 'ones' or 'zeroes'. The latter allows bandpass filters which are about as narrow as 50% of the span between the first two 'nulls'. GPS receivers prefer filters that are 2 to 3 times wider than the distance between the two 'nulls'. See Figure 9.1.

High-end GPS receivers can maybe upgraded accordingly the JAVAD approach. Unfortunately, this method is nearly impossible to implement in

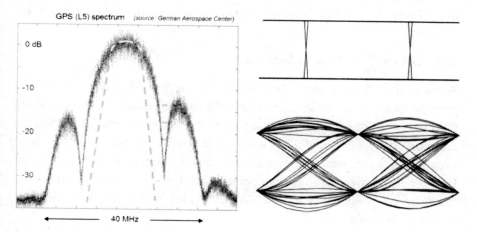

Fig. 9.1 The L5 GPS spectrum (blue) as sent is approximately 40 MHz wide. If this code would be used for telecom applications, the spectrum could be reduced down to less than 20 MHz without introducing data errors. The upper eye pattern on the right show the steep slopes needed to achieve well-defined correlation peaks. The lower eye pattern is indicative for telecom data transmissions. Although the bandwidth is significantly reduced as shown by the orange line in the spectrum plot, the sloppy slopes do not endanger the data detection reliability.

the millions of GPS receivers used in smart phones and car navigation systems. Should all existing GPS C/A type receivers simply be depreciated and destroyed? And who will pay for this enormous logistically challenging modification or retro-fitting process? We may expect an interesting series of law suits.

An underlying issue is the commercial spectrum business which is for GPS and for telecom so different. GPS and all other GNSS systems have free access to the used 60 MHz wide spectrum while the telecom providers usually have to pay for this. For example, the G3 30 MHz wide spectrum in the UK has been sold by auction for 1.350 billion Euros/year which is 22 Euro/year/capita. An identical approach happened in the Netherlands where the Dutch pay about 9 Euro/year/capita. These auctions yield relatively easy incomes for governments as the tax mechanism is simple. If this should also be done with the use of the GNSS bands, how to collect these taxes, and which government would be allowed to collect this money? Should the GNSS providers be authorized for this, and would, for example, South American users want to pay for US, Russian, Chinese or European spectrum claims? Who owns the spectrum anyway? Can anybody sell something which he doesn't own? This lack of balance between the GPS community and the telecom providers may disturb the relations between the two large and powerful communities.

9.6 Steps Towards a Solution

A solution shall be found as both, GPS and telecom networks are essential parts of today's economies. GPS might shrink their spectrum needs and the GPS industry/providers/users could start to pay for spectrum. Both solutions are most likely just wishful thinking, so, the only remaining solution is to improve GPS receivers so they can withstand the new powerful neighbours. However, this may cause a costly operation and logistically nearly impossible to realize in a couple of years. But, JAVAD made a first step which is difficult to ignore! Who's next?

9.7 Conclusions

- Pressure on the GNSS spectrum will not just continue but will even increase in many parts of the world.

- GPS spectral claims are primarily based on current technology and on its enormous installed user base.
- Spectrum is becoming scarce, so the price will rise but the telecom industry shows that this does not block the business profitability.
- Balancing the 'spectrum rent' costs between GPS and telecom users might help to relax the discussions.

Biography

Professor Dr. Durk van Willigen (1934) retired as appointed professor in Electronic Navigation Systems at the Delft University of Technology in The Netherlands where he headed a group of researchers and students for 11 years. His group was very active in the field of GPS receiver signal processing algorithms to reject multipath (MEDLL) and interference. In the area of advanced guidance and navigation displays, this group was heavily involved in several national and international projects focusing on the development and evaluation of so-called synthetic vision systems.

Dr. Van Willigen initialized Eurofix in 1989 and has been the principal investigator since. He has published many papers on navigation systems in general, and on Loran-C and Eurofix in particular.

He is the recipient of the Thurlow Award 1999 of the United States' Institute of Navigation, and of the Gold Medal 2002 of the Royal Institute of Navigation in the UK. The International Loran Association awarded Dr. van Willigen the Medal of Merit in 1996 and the John M. Beukers Award for Technical Innovation in 2005.

Professor Durk van Willigen founded Reelektronika in 1975; a consulting company specialized in radio positioning systems, navigation, radar, and signal processing for land and marine applications.

Dr. Van Willigen is Chairman of the Board of the GAUSS Research Foundation and member of the Advisory Board of the Netherlands Institute of Navigation.

Conasense Charter: "Hollistic Exploitation of the Electromagnetic Spectrum"

Ole Lauridsen

Aalborg University, Denmark

The existence and utilization of electromagnetic waves, first predicted by Maxwell, later proven by Helmholz and finally exploited by Marconi, who triumphed in 1902 by spanning the Atlantic between Europe and United States, is a basic condition of the world and space, as we know it of today.

COmmunication, Navigation and SENSING are the three main disciplines of "radio technology".

In the last century ways of sharing the access and use of the electromagnetic spectrum has become a source of big frustrations and despair. After the Berlin radio conference in 1903 different frequencies and time of the day became the tools for giving access to many radio users.

At the start radio use was primarily for naval-, military- and governmental applications. Parallel to this, a society of radio amateurs grew up. The catastrophe of the Titanic, 1912 and the First World War were incidents that spurred the technology for better and more secure radio communications. The large number of powerful transmitters soon gave way for Radio navigation by sensing of known transmitters and determining the direction to them.

COmmunications- NAvigation-SENsing-SErvices (CONASENSE), 191–192.

From 1934 to 1939, most European countries planned for using the 40 to 50 MHz band for local modern high quality audio broadcast including FM and Television. The upper edge was chosen by the fact that this was the limit for normal vacuum tubes and coil technology. The technology development during World War 2 has led to new radio technology covering frequency bands up to 10 GHz, while the advent of navigation aids and radar led into a post war scenario with a huge number of new radio based services all crying for more and new spectrum. The present day situation with Space satellite communication and navigation, GPS, Glonass and others has together with the wish of undisturbed Stellar radio astronomy, not made the situation easier in a world with more than 3 billion GSM personal radios!

Essence for CONASENSE: How to share and exploit the spectrum

Methods of spectrum use have always been based on dividing according the needed frequency bands and type of service. When we divide by frequency band, we always loose up to half of the spectrum in inaccuracies of filters and oscillators. Many navigation systems are serving users who also need communication, so why not look into a much more intelligent way of using the technology and spectrum by combining communication, navigation and sensing applications in a more intelligent way to offer the populations of the earth a better and clever exploitation of the electromagnetic spectrum to the benefit of mankind!

Of course CONASENSE will result in doing outstanding research in generating, receiving and analyzing radio waves.

Index